A NATIONAL STRATEGY FOR IMPROVING PRODUCTIVITY IN BUILDING AND CONSTRUCTION

Proceedings of the
Building Research Advisory Board's
1979 Building Futures Forum, Washington, D.C., 1979.
November 7-8, 1979, Washington, D.C.

Organized by the
Technology Assessment and Utilization Committee
Building Research Advisory Board
Commission on Sociotechnical Systems
National Research Council

NATIONAL ACADEMY OF SCIENCES
Washington, D.C.
1980

NOTICE: The project that is the subject of this report was approved by the Governing Board of the National Research Council, whose members are drawn from the Councils of the National Academy of Sciences, the National Academy of Engineering, and the Institute of Medicine. The members of the Committee responsible for the report were chosen for their special competences and with regard for appropriate balance.

This report has been reviewed by a group other than the authors according to procedures approved by a Report Review Committee consisting of members of the National Academy of Sciences, the National Academy of Engineering, and the Institute of Medicine.

Library of Congress Catalog Card Number 80-81951

International Standard Book Number 0-309-03080-3

Available from

Office of Publications
National Academy of Sciences
2101 Constitution Avenue, N.W.
Washington, D.C. 20418

Printed in the United States of America

BUILDING RESEARCH ADVISORY BOARD

THE BUILDING RESEARCH ADVISORY BOARD is the principal unit of the National Research Council Commission on Sociotechnical Systems concerned with the built environment. Its primary objective is to promote the orderly growth and development of science and technology to meet societal needs for housing, building, and related community and environmental design and development. In fulfilling this objective the Board works toward: improving the availability and utilization of the human and natural resources needed to create, operate, maintain, and renew the built environment; identifying the individual, societal, and physical environmental objectives to be sought in the built environment; increasing the effectiveness of the building process functions of planning, design, construction, operation and maintenance, repair and alteration, rehabilitation and disposal, and recycling of useful materials and products; and extending the base of knowledge concerning the art, science, and technology of building and stimulating the transference of this knowledge into application.

The 24-member Board, together with a number of individuals from federal agencies who serve in a liaison capacity, constitutes a body of recognized authorities in building-interested segments of industry, government, the design professions, and academic and research institutions; each member serves as an individual and never as a representative of any other organization or interest group. The Board's strength rests in the expertise and dedication of this multidisciplinary membership and that of the hundreds of individuals from both the public and private sectors who respond to its requests that they serve voluntarily on its various working groups. Composed of these knowledgeable individuals, the Board's Councils, committees, panels, and task groups address and often recommend solutions to important problems and issues that affect the future of the built environment.

BRAB TECHNOLOGY ASSESSMENT AND UTILIZATION COMMITTEE

Chairman

DAN E. MORGENROTH, Director, Market Development, Owens-Corning Fiberglas Corporation, Toledo, Ohio

Forum Program Chairman

MAURICE H. GARDNER, Manager, Projects Operations, General Electric Company, Schenectady, New York

Committee and Subcommittee Members

GIFFORD H. Albright, Head, Department of Architectural Engineering, The Pennsylvania State University, University Park
JOHN J. ANDREWS, PE, Vice President, Smith Hinchman & Grylls Associates, Inc., Detroit, Michigan
WILLIAM S. BIRNEY, Marketing Manager, U.S. Steel Corporation, Washington, D.C.
JACK M. CHERNE, PE, Manager, TRW Energy Systems, Redondo Beach, California
G. DAY DING, Head, Department of Architecture, University of Illinois, Urbana
PAUL C. GREINER, Vice President, Conservation and Energy Management, Edison Electric Institute, Washington, D.C.
SAMUEL L. HACK, Management Consultant, Washington, D.C.
CALVIN S. HAMILTON, Director of Planning, Department of City Planning, Los Angeles, California
DAVID S. HAVILAND, Associate Dean and Director, Center for Architectural Research, Rensselaer Polytechnic Institute, Troy, New York
CARL HOUSE, Consultant, Washington, D.C.
OLIVER H. JONES, Consulting Economist, Oliver Jones and Associates, Manns Choice, Pennsylvania
GLENN H. LOVIN, Manager, Alternate Energy Applications, Edison Electric Institute, Washington, D.C.

THOMAS J. LUCAS, Jr., AIA, President, TMP Associates, Inc., Bloomfield Hills, Michigan

DONALD LYON, Engineering Manager, Design and Construction, American Telephone and Telegraph Company, Basking Ridge, New Jersey

FRANK J. MATZKE, FAIA, Vice President, Technology and Programs, National Institute of Building Sciences, Washington, D.C.

H. ALLYN PARMENTER, Training Department, United Association of Journeymen and Apprentices of the Plumbing and Pipe Fitting Industry, Washington, D.C.

DONALD SAYLAK, Associate Professor, Mechanics and Materials Research Center, Texas A&M University, College Station

AARON SABGHIR, Director, Construction and Building Products Division, Bureau of Domestic Business Development, U.S. Department of Commerce, Washington, D.C.

HILARY W. SZYMANOWSKI, Manager, Advanced Conversion Technology, Westinghouse Electric Corporation, Pittsburgh, Pennsylvania

WARREN H. TURNER, Engineering Director, Equipment and Building, American Telephone and Telegraph Company, Basking Ridge, New Jersey

BRAB Staff

JOAN D. FINCH, Executive Secretary, Technology Assessment and Utilization Committee

CLARET M. HEIDER, Editorial Consultant

M. S. BLACKISTONE, Consultant

SPONSORS OF THE 1979 BRAB BUILDING FUTURES FORUM

Public Sponsors

NATIONAL SCIENCE FOUNDATION
U.S. DEPARTMENT OF COMMERCE
U.S. DEPARTMENT OF LABOR
U.S. DEPARTMENT OF TRANSPORTATION, URBAN MASS TRANSIT ADMINISTRATION

Private Sponsors

AMERICAN TELEPHONE AND TELEGRAPH COMPANY
BUILDING AND CONSTRUCTION TRADES DEPARTMENT OF THE AFL-CIO
EDISON ELECTRIC INSTITUTE
FEDERAL NATIONAL MORTGAGE ASSOCIATION
FILLEY AND COMPANY, INC.
GENERAL ELECTRIC COMPANY
JOHNS-MANVILLE CORPORATION
NATIONAL ASSOCIATION OF MUTUAL SAVINGS BANKS
OWENS-CORNING FIBERGLAS CORPORATION
UNITED STATES STEEL CORPORATION

PREFACE

The thought-provoking presentations by the distinguished participants in BRAB's 1979 Building Futures Forum successfully focused attention on the issues of productivity in the construction industry. BRAB extends its thanks to the people who organized this Forum and to the members of the industry who participated.

Conversations with the various attendees indicate that the Forum gave them a very different perspective on the true magnitude and nature of problems relative to productivity in construction. Many participants who began with concerns relative to a narrow interest area such as financing, regulations, or worker training broadened their understanding of the factors that constrain productivity.

From Dr. Siegel's keynote address until the conclusion of the last panel, it was evident that the national strategy for productivity in the construction industry will not come from Washington. Good life cannot be legislated. Instead, it will take a concerted effort and solid support from the entire building and construction community at all levels to improve productivity.

Productivity is inhibited by the rules that have grown up around us with time. We will have to learn to break the bonds of tradition and to mobilize all of our individual interests for the good of the industry. We realize there are no simplistic solutions to improving productivity in the construction industry.

As a result of this Forum, the BRAB and its TAU Committee organization will begin efforts to develop realistic programs directed toward solving problems concerning productivity. We intend to select a limited number of priority projects and execute them well.

Finally, I would like to quote the chairman of the Board of the Committee for Economic Development, Fletcher Byrom, who sums up the nature of the problem: "America has a proud record of world economic leadership. We have not lost our lead, but we have hobbled ourselves unnecessarily in recent decades and are hurting from the consequences. We do not have to look elsewhere for blame or for solutions. We got ourselves into this bind. We have the knowledge and the power to get ourselves out of it. What matters is whether we have the will."

Support of and participation in future programs of the BRAB will help lead us to a new age of cooperation and efficiency in the construction industry.

Dan E. Morgenroth, Chairman
BRAB Technology Assessment and
Utilization Committee

CONTENTS

Session III
GOVERNMENT AND PUBLIC ACTION FOR IMPROVING PRODUCTIVITY

Session IV
HUMAN MOTIVATION AND INCENTIVES FOR IMPROVING PRODUCTIVITY

Session V
INNOVATION FOR IMPROVED PRODUCTIVITY

INTRODUCTION

The Building Research Advisory Board (BRAB) convened its 1979 Building Futures Forum in November 1979 to focus on a national strategy for improving productivity in building and construction. The Board recognized that demands placed on the building and construction community have changed dramatically in the past decade. In response to these demands, all segments of the construction industry are seeking ways to achieve greater efficiency in meeting their own needs and the needs of society despite changing availability of resources, rising costs, increasing regulation, and conflicting priorities.

The Forum brought together representatives of all elements involved in or influencing construction. The primary objective of the Forum was to provide a foundation for improving productivity by:

1. Delineating and developing a better understanding of the factors influencing productivity in the built environment,
2. Determining how these factors affect productivity,
3. Determining what must be done to improve industry efforts, and
4. Making recommendations and identifying appropriate responsibility for necessary action.

In planning for the 1979 Forum, the BRAB Technology Assessment and Utilization (TAU) Committee conducted several planning sessions to assure that the Forum would meet its primary objectives. Three of these were workshops at which 65 experts from all fields of building and construction were asked to raise critical issues revolving around the productivity question. It was found that many of the issues identified reflect traditional building and construction industry problems. Thus, the challenge to the Forum participants was to become active in forming a united industry to stimulate actions that will enhance the productivity of all by solving these problems.

The Forum was organized into five major sessions:

1. Managing the building process for improved productivity,
2. Financial planning for improving productivity,
3. Government and public action for improving productivity,
4. Human motivation and incentives for improving productivity, and
5. Innovation for improved productivity.

During each session, the session chairman presented an overview of the critical issues and problems identified in the planning workshops in his particular subject area. A panel of experts then addressed the issues and potential solutions to the problem and responded to audience questions. In these proceedings, the welcoming address, keynote address, and forum overview are followed by summaries of the five sessions. The statements of the panelists and the discussions that concluded each session (except the last) then are presented. The Forum participants are listed in appendix A and biographies of the speakers are presented in appendix B.

WELCOMING ADDRESS

ROBERT M. WHITE
Administrator
National Research Council, Washington, D.C.

It is a very great pleasure for me to welcome you on behalf of the National Academy of Sciences and the National Research Council. The 1979 Building Futures Forum will address the possibility of developing a national strategy for improving productivity in building and construction, a central issue of our industrial society.

As the Forum outline attests, improving productivity is not a problem of management alone or of finances, research and development, governmental action, or cultural attitudes. Indeed, it involves all of them.

I am reminded of the ancient Greek tale of Laocoon, who struggled to free himself from serpents, and of the Japanese film "Rashamon," in which the witnesses of an incident each described it quite differently. So it is with the issue of productivity. Disparate actions in one sphere of society or another, although individually small factors, coalesce into a mighty impact whose net effect results in the dilemma to be addressed by this Forum. It is no wonder, then, that we struggle to understand and try to find solutions to this problem of productivity.

But that is your task. My task is to tell you something of this institution and your surroundings. Some of you may be new to the Academy complex. It is composed of the National Academy of Sciences, the National Academy of Engineering, the Institute of Medicine, and the National Research Council, their common operating arm. It is an institution with great traditions in the service of the nation. It was founded in the administration of President Lincoln in 1863 to provide scientific and technical advice to the government. In 1916 during World War I, the National Research Council became the operating arm of the National Academy of Sciences. The National Academy of Engineering and the Institute of Medicine are newer entities of this Academy complex reflecting the increasing needs of the federal government for specialized help in these fields.

There is hardly an area of governmental interest in which the National Research Council is not active. None of these, however, is more important to the long-term welfare of our society than the issue of productivity. It is with great enthusiasm, therefore, that we welcome the initiative of the Building Research Advisory Board's (BRAB)

Technology Assessment and Utilization (TAU) Committee in sponsoring this Forum. Joseph Zettel, Chairman of BRAB, Dan Morgenroth, Chairman of the TAU Committee, and Maurice Gardner, Chairman of the Forum Program Committee, are to be congratulated for their fine work.

KEYNOTE ADDRESS

IRVING H. SIEGEL
Consulting Economist
Bethesda, Maryland

Not being asked often to serve as a keynote speaker, I have come prepared to make remarks that are actually relevant to the present conference. These remarks focus on two principal themes. One is the chronic weakness of the statistical base of the construction industry, particularly in the productivity area. The second concerns the practical matter of improving construction productivity--an enterprise that does not critically depend on the upgrading of statistics or on any new federal policy initiative.

I began work in the productivity area in 1936, and, by coincidence, my first assignment was to help design and conduct a survey of productivity in the construction industry. This activity was undertaken as part of the Works Progress Administration's (WPA) National Research Project on Reemployment Opportunities and Recent Changes in Industrial Techniques. At its peak, as I have jokingly told many audiences, this WPA project employed half of the nation's unemployed intellectuals in trying to find out why the other half were not working.

This initial productivity venture taught many lessons that have remained regrettably valid. First in importance was the failure of the study to reach completion for a reason that has since become too familiar--understaffing from the start and switch of resources in midcourse to other worthy studies that seemed to promise earlier payoff at lower cost. Second was the paucity or inaccessibility of needed data. Third was the difficulty of measuring construction output and productivity, whatever the condition of the data.

The WPA approach to measurement had an element of novelty that still merits imitation. Well defined types of buildings (e.g., schools and warehouses) were selected for study, and an attempt was made to locate and visit contractors (if not subcontractors) with records for more than one structure. The buildings were divided into component "units of work"--excavation, foundation, frame, exterior walls, interior walls, floor, roof, plumbing and heating, etc. For each of these units (which I call "subproducts" in contrast to the more heterogeneous "end products" that productivity statisticians conventionally measure), we sought to match quantities of output with man-hours of input. From the resulting ratios, we intended to derive man-hour estimates (with differentiation according to skill or craft) for a standard synthetic structure over the years. Although I am aware of subsequent studies of productivity for various crafts, I know of none that has attempted to track changes in labor requirements for a standardized complete structure.

Turning to the larger scene, I find the state of production and productivity measurement for construction still disappointing. I say "still" because integrated data on output and input are not yet compiled, by type of construction, on a national basis--and are not soon expected to be. This prospect may seem unreasonable in view of the size of the industry. The Forum brochure tells us that the industry accounts for about 10 percent of the gross national product, for direct employment of about 4 million persons, and for employment of about 3 million more who supply auxiliary services and materials. This prospect also may seem unreasonable in view of the frequent complaints made over the years concerning the inadequacy of the statistical base, the pitfalls of the method of price "deflation," and the technical feasibility of measurement in terms of "units of work" or "subproducts." Extenuating factors can, of course, be cited. The industry is, after all, diffused, variegated, and responsive to all sorts of customer needs--a remarkably adaptive, protean system that can profitably assemble funds, people, equipment, and materials for the realization of wanted structures. Such an organismic view of the industry and its accomplishment is attractive, but it could encourage complacency with regard to the deficient data base.

I recall numerous occasions on which the egregious flaws of construction statistics have been noted. In 1948, a report of the Joint Economic Committee of the Congress, entitled Statistical Gaps, cited, among other things, a need for outlays of about $5 million per year (about $25 million in today's federal salaries) to remedy defects in the construction field. Similar defects were mentioned in a paper presented at a 1972 conference sponsored by the National Commission on Productivity. The 1976 BLS Handbook of Methods warned again of the inadequacy of output, productivity, and cost statistics for the construction industry. This year, a National Academy of Sciences panel reviewing productivity statistics is scheduled to release its final report, which will once more refer to the weakness of output and productivity measures for construction.

Although our chief interest here is in productivity, other limitations of the data base for construction should not escape notice. The difficulty of devising suitable price deflators for estimating the volume of output from value figures reminds of the upward thrust imparted to the Consumer Price Index by the treatment of home prices and interest rates during this prolonged period of inflation. Speaking of home prices, I have been impressed, as a consultant, that there is no index of U.S. land prices--a surprising gap when you consider that land now accounts for about 25 percent of the price of a home. In 1949, according to the National Association of Home Builders, the land share was only 11 percent. (Congressman Reuss, I understand, is seeking to have this data gap plugged.)

Since I belong to the small minority of productivity students that openly doubts the reliability of most measures derived via price deflation during a prolonged inflation, I shall not lean heavily on the numbers available for construction. As you probably know, a slowdown is indicated in official and unofficial estimates of output per man-hour for the economy at large, especially since 1973. The rate of change for construction in recent years is weakly positive or even negative. You may wish to consult the 1979 Economic Report of the President to get an idea of the computed changes for construction and the other major sectors. Another set of figures appears in Survey of Current Business, August 1979, Part II. Whether the numbers are accepted or not, they say nothing, of course, about what you may expect on a particular job, at the microlevel. At the worksite, management seeks to ensure that all costs, labor and others, are as small as possible while also as large as necessary, and productivity is served thereby even without explicit measurement. That productivity in construction as a whole also could be improved without explicit measurement is another matter about which I shall say something shortly.

While still on the subject of actual productivity statistics, I want to alert you to a possible unwholesome productivity improvement on the national scene as the recession, so dear to many economists, becomes a reality. Despite economic wiggles, employment has held firm in recent years. If large doses of monetary hemlock succeed in purging speculative fever on the price front, they also may purge speculative employment. If expectations of the business community change drastically, policymakers may be rewarded with a flood of unemployment rather than with a gentle rain. The numerator then may increase with respect to a rapidly declining denominator--the wrong kind of productivity gain, one that does not foretoken higher living scales. Indeed, even bank economists and others who, driven beyond wit's end, have opted for a recessionary cure of inflation may, in these circumstances, find their own jobs endangered.

Before proceeding to my second main theme, I want to offer another forward-looking observation. Dramatic interest-rate increases coupled with recession set the stage for a major shift in the relationship between consumption and investment. This shift is probably already under way, and it may well be confirmed by "unpopular" deliberate changes in the tax system that support capital formation. The mixed implications for construction will not be explored here.

The Forum brochure provides an entry point into my second topic. It states that this Building Futures Forum is devoted to developing a national strategy for improving productivity in building and construction and that you are to go home and take actions that will improve the service and the productivity of the industry. You are challenged to work toward these ends as members of a united building and construction community. "Strategy" and "community" are key words. Is there some way to translate them from the rhetoric of a conference brochure into living reality? Let us be naive enough to address this issue head-on.

In recent years our country seems to have moved far toward embrace of a notion ridiculed in a great line of T.S. Eliot in the 1930s--the notion that we can have "systems so perfect that no one will need to be good." Systems certainly can be designed to compensate for human lacks in many respects and to varying degrees but, in the last analysis, even the functionality of these systems depends on the presence or absence of a personal quality that has been ignored or taken for granted in the design. During this conference and especially in the fourth session on human motivation and incentives for improving productivity, you surely will hear of the role of "human resources." To me, this has always been a depressing term, conjuring up images of inert people waiting to be molded or cannibalized, of people as property or chattel. Can we change the term and improve the concept by speaking instead of "resourceful humans"? Can this dynamization be effected in a system not yet designed to encourage people to be "good"? Can we motivate people to reprogram themselves to be "good" without regard to the requirements and constraints of the system in which they normally operate? In short, may we realistically expect participants in a conference, this conference, to go home and act as "resourceful humans"? This is the challenge of the Forum brochure, perhaps disconcertingly restated.

I literally believe that commitment to act as a "resourceful human" on your home terrain is the critical ingredient, at the present time, of a national strategy for improving productivity in construction. It is not necessary to wait for federal initiative and funding, for design of a grand, comprehensive "system" in which your town has its assigned part. Instead, start the "strategy," if there will ever be one, by being "good" at the grass roots, by doing spontaneously and better what you already know is needed. Let a "national strategy" evolve through aggregation. Exercise primary leadership in your communities; then come to Washington later for supportive or reinforcing policy.

Am I merely indulging in a substitute rhetoric? I think not. I am saying what is eminently "practical," and I shall document with cases. I shall cite impressive evidence contained in an unpublished report, dated September 1978, of the defunct National Center for Productivity and Quality of Working Life. By the way, the very demise of this organization at a time of heightened public concern over our productivity performance underscores the importance of doing what can be done at the local level without awaiting a signal from an uncertain federal trumpet.

The unpublished report (prepared by Louis Alfeld with assistance from Joseph Russell on the basis of material developed by Quinn Mills) describes six geographic areas in which construction unions and management have sought to nurture community. In "memoranda of understanding," labor and management have set forth rules for improving productivity to their mutual benefit, to reduce the obstacles and attenuate the antagonisms that impede timely and economical construction. As I drafted my notes for today's meeting, I labeled this section "From Acrimony to Acronym." In a moment you will see why.

In one of the six places, St. Louis, the program to implement the joint labor-management compact is called PRIDE (Productivity and Responsibility Increase Development and Employment). In Reno, the program is SIR (Skill, Integrity, and Responsibility). The Columbus program is MOST (Management and Organized Labor Striving Together). When we get to Boston, we have run out of acronyms and are left with an incorrect Roman numeral that still stands for a worthy idea: MLMCC (Massachusetts Labor and Management Construction Committee). The Colorado program is called Union Jack and the Indianapolis endeavor, Operation Topnotch.

In none of these six instances was federal funding, prompting, or participation a factor. The joint labor-management councils have sprung up to deal with outdated work rules, jurisdictional disputes, work attitudes, starting times, and other matters impinging on productivity because a threat to the common interest is perceived. On the local scene, union-oriented contractors and workers have to reckon with the potential non-union competition. How pleasant it is to record a response to an economic threat that favors, rather than hinders, productivity.

Some additional evidence on the formation of self-help configurations has come to my attention. For example, in 1975, the Oregon-Columbia Chapter of the Associated General Contractors signed a five-year agreement with a group of unions that established a "multicraft productivity board." Obviously, much can be done by local leaders who determine to be "good" on their home grounds without the benefit of federal prosthesis, without the guidance of a pre-established "national strategy." True, a national voice, a national beacon could help the quest for mutuality, but, for the time being, the Building Research Advisory Board and national employer and labor organizations should be able to provide some of the needed coordination.

In preparing for today's appearance, I looked through a 2-foot pile of material, mostly workshop minutes and papers that BRAB kindly provided. On reading through this material, I found myself nodding my head in continual agreement. During my 43 years of interest in productivity, I had encountered the same thoughts often. So much of what I read made sense. But what should at last be done? I am reminded of a constant query from practical businessmen who have heard me lecture on productivity: "Doc, let's get down to the bottom line. What's the single most important thing in productivity?" I invariably snap back: "Whatever you have left out."

What seems to be left out in the construction industry (as elsewhere) is the unprogrammable behavior of the "resourceful humans" who will insist on doing what needs to be done to make a system work. Such people obviously exist throughout the country. They are not waiting for a perfect system, a "national strategy" centrally designed and administered. Their ranks should be swelled by attendees at conferences such as this.

FORUM OVERVIEW

C. E. PECK
Executive Vice President
Owens Corning Fiberglas Corporation, Toledo, Ohio

This is a Building Futures Forum. We will be talking about future actions--about your individual actions as well as construction industry actions and government actions. Improved productivity in construction, however, will depend mostly on what we can do as individuals.

Better productivity in construction is certainly important in meeting society's needs and desires for a better built environment. The product of construction is indispensable for human survival, because it provides both shelter itself and the better life. Over 270 recognized building types attempt to provide the proper environment for diverse activities that range from family life to manufacturing and from worship to athletics. Shortcomings of the building community in areas such as rising costs, unemployment, or supposedly low productivity have a heavy impact on the total economy. When the cost or the availability of homes for American families is affected, society as a whole is going to get involved in construction industry problems. Thus, BRAB's singling out of the issue of productivity in the construction industry for study and public discussion in this Forum and its publication of the concepts and facts reviewed here is very, very appropriate.

In this Forum, we would like you to consider the word "productivity" to include in its scope the widest possible range of improvements in efficiencies or in utilization of resources. We should not confine ourselves to the narrow definition of the economist. Improved productivity relates to <u>all</u> the things that you can do, that we can do, that the industry as a whole can do better. It includes the productivity of investment, of money, and of land as well as of labor.

Let us be sure that we review the meaningless work that we impose on buildings just as much as we review our efficiencies in doing the necessary work. Let us be sure that we consider the wastes of poor management just as much as we consider the wastes of poor labor. Let us be sure that we recognize the productivity drag of the regulatory and approval process. Let us be sure that we are concerned about the poor productivity of capital that is implicit in the stretched-out time periods required for construction in today's social environment.

There are five unique characteristics of the construction industry that constrain our ability to be productive. First, construction is, by its nature, immobile. It stays where it is built. This fact leads to designs and concepts that really aim at utilizing the unique opportunities presented by the land and the location as much as by the structure or the building itself. This factor directs us away from standardization of the product as being the optimum economic thing for the best total result. Second, geographic dispersion is the nature of construction needs. I think one of the wonders of our construction industry is that in all of the thousands of counties across the nation, you can get any kind of building built that you might imagine and get it built at a reasonable cost and in a reasonable period of time. Geographic dispersion has led to a relative labor intensity of the process of building as compared to other industries. Third, the industry has many, many different specialized segments, and it is served by a large number of small independent businessmen. Fourth, the "boom and bust" business cycle of construction has led to a turnover of businesses in construction that is one and a half to two times higher than the turnover in the manufacturing industry. As a result, we are constantly re-recruiting and retraining our labor. Perhaps more importantly, we are leading much of that labor with new untrained managers. Because more buildings are built during the "boom" cycle, this wastefulness is highly inflationary. Fifth, research and development expenditures in construction do lag behind those of other large industries.

These five elements are fundamental to our industry and we need to recognize them; nevertheless, we can improve productivity and we will if we want to and aim at this improvement as a goal. Progress will begin with individual initiative. Momentum will come from broader adoption of ideas, and you will hear many good, solid ideas as you listen to the Panels in this Forum. Full speed will follow if we can achieve some structural changes in the industry over the coming years.

To define the focus of this Forum, three planning workshops were held early in 1979. Sixty-five industry leaders representing building owners, building operators, investors, economists, designers, planners, manufacturers, suppliers, builders, contractors, labor, government, and public interest groups were brought together in these planning workshops to identify the critical issues that comprise the productivity problem. These workshops covered planning and management's role, technology's role, and the public's role. Many similar concerns emerged from each of the workshops. Five common themes were discovered: managing the building process, financial planning, government and public action, human motivation and incentives, and innovation. Each of these themes will be discussed by a panel during the Forum sessions.

What we are seeking through this Forum is a united effort to come to grips with the question: How can we build and operate more efficiently with available resources to provide the buildings that society demands and needs? The Building Research Advisory Board solicits your participation in recommending and implementing actions that will keep our industry strong and healthy in the years to come. There is no question about it; we <u>can</u> improve productivity.

SESSION SUMMARIES

SESSION I: MANAGING THE BUILDING PROCESS FOR IMPROVED PRODUCTIVITY

SAMUEL L. HACK
Management Consultant
Washington, D.C.

In examining management's role in improving productivity in the building process, the panelists emphasized repeatedly that the role and responsibility begins with the owner, the developer, or the developer-manager. He should delegate responsibilities, in accordance with the scope, schedule and budget of the project, to monitor progress and identify deviations from the plan and to control all aspects of the project and the process. A major problem, however, is that many owners do not appreciate their role and responsibilities in the process. Many assume that the award of contracts and the delegation of responsibility under these contracts relieves them of responsibility. They do not recognize the need for continuous interplay in project execution. Although many owners are very sophisticated participants in the process, many others do not know their options, what expertise is available, what their various contract approaches can be, or what practical technical, cost and schedule goals can be set.

In view of these problems, the Panel felt that management can make the greatest contribution to improved productivity by developing a strategy that would produce a construction program of the shortest duration. The goal is to get in and out as quickly as possible. Communication is an essential part of such a strategy since each participant must clearly understand his responsibilities and authority and the responsibilities and authority of the other participants. The aim of this approach is to stimulate a team rather than an adversary relationship.

More specifically, the Panel believes it is necessary to develop for and distribute to owners and managers guidelines or checklists of factors to be considered in developing a management plan for a construction project. The question of how to contact the uninitiated owner or manager remains unanswered. One possible method would be to utilize lending institutions or funding groups to educate owners and managers in what should be done at the time they are proceeding with a project.

The Panel also concluded that there is a need to create a new organization or strengthen an existing one to serve as the focal point of the industry's advocacy. The general consensus was that the construction industry is not an industry in the same way that the automobile manufacturers or the oil producers are. Finally, the Panel believes that there is a shortage of broadly skilled individuals who can serve effectively as project managers and that it is essential to develop such individuals, either through academic or on-the-job training.

SESSION II: FINANCIAL PLANNING FOR IMPROVING PRODUCTIVITY

AARON SABGHIR
Director, Construction and Building Products Division
U.S. Department of Commerce, Washington, D.C.

The challenge of the financial planning session was posed in terms of how money or financing can be used to improve physical productivity. When considering the subject at the macro level, it was concluded that money problems arise because the building community must complete in capital markets with other economic sectors and because monetary and fiscal policies result in uneven flow and availability of this resource. Thus, the cyclical nature of the economy affects construction more severely than it does other sectors of the economy. The inadequate rate of long-term savings in the United States also was emphasized as creating problems as were factors such as the role of taxes and the corrosive effect of inflation, which has tended to discourage savings in the United States.

Productivity problems at the micro level result because many firms have inadequate equity capital and are subject to sharp seasonality factors. The Panel concluded that there is a lack of perspective in regard to using capital investment in construction to promote life-cycle economies. All of this seems to reflect a lack of financial policy and planning at both the macro and micro levels.

The Panel emphasized that there is a lack of predictability of the cost of money, particularly in terms of looking ahead two or three years. This, of course, is accentuated by the business cycle and, in short, the uncertainty element that hinders attempts to achieve good management or sound market analysis. Even the development of many new market instruments and sources of money, it was pointed out, has not been sufficient to alleviate the uncertainty aspect of the money problem.

The inevitability of cycles and the crucial relationship between equity and debt capital were discussed. The need to learn to live with the cycle by building up equity in good times was identified as an important objective. The deliberate practice of financial management by the building firm was viewed as very essential, and in that connection, the use of joint ventures with both manufacturers and other contractors was cited as offering potential for strengthening financial resources in the construction industry. In other words, there is a need for a positive approach to the management of money flows and the projection of money needs for both individual projects and the more general overall operations of the construction industry.

Finally, the need for such things as tax credits and the elimination of arbitrary rules on interest rates was mentioned repeatedly. The need to remove Federal Housing Administration and Veterans Administration rate ceilings in the housing sector and the need for a pre-emptive national law instead of state usury laws were cited. The Panel also concluded that new types of mortgage instruments and approaches to financing are needed. Such instruments as graduated payment and rollover mortgages were referred to as a possible reflection of future trends. In short, the Panel looks to both the government and the private sector to develop the many things that are needed to make the use of money and financing a more powerful instrument for improving physical productivity.

SESSION III: GOVERNMENT AND PUBLIC ACTION FOR IMPROVING PRODUCTIVITY

FRANK J. MATZKE
Vice President, Technology and Programs
National Institute of Building Sciences, Washington, D.C.

The Panel focused primarily on government and public actions that have an impact on productivity. Some examples were given of government actions, particularly at a local level, that facilitate the approval process. It was emphasized, however, that public policy at all levels, as expressed in regulations, generally produces a negative impact on the building process in the sense that most regulations add to the time that it takes to do the work and to the cost of the work. Little was said about the positive benefits of public policy as expressed in regulations. The chairman, however, indicated that the public should be informed of the cost of requirements. When policy is expressed in the form of regulations, the value of the benefits to be gained should be equal to or greater than the cost of the regulation.

Another thought expressed during the Panel's discussions was that federal regulations often generate problems that go beyond the intent and purpose of a particular regulation. For example, environmental quality legislation like the Clean Air Act in effect leads to restricted land use.

Several long- and short-term recommendations for improving productivity in relationship to the regulatory process were made. One short-term recommendation is to follow the model of the city of Detroit, which has established a one-stop permit approval system. The Panel believes that this approach could be carried forth in other areas and at other levels. Eventually, it might be possible to obtain approvals of plans and specifications for new construction at the local level that would meet all requirements--federal, state, and local.

Over the longer term, the Panel concluded that there is a need to relate regulations to the proper level of concern and to administer them at the appropriate level. For instance, there are federal, state, regional, and local regulations that impact in one way or another on the building process. Somehow these must be sorted out. What is required for the national good should be expressed in federal regulations. What needs to be done that relates more to state requirements and regional requirements should be handled at those levels. What needs to be done to satisfy unique local requirements should be handled at that level.

It was suggested that the next step would be to pass on authority or responsibility to the local level so things could be handled in one place. Although that would involve a long-term effort, there are ways that it could be done. One of the panelists presented a proposal for allocating responsibilities to the logical level of government. This would involve: consolidating responsibilities wherever possible and, in that way, avoiding duplication; simplifying procedures; reducing the time required for review and approvals; placing responsibility at the grass roots level whenever possible; instituting "sunset" review procedures at each level of government to ensure that unnecessary regulations do not remain in force; and crosschecking legislative proposals with basic goals and objectives before a regulation is actually finalized.

To date, elements of the building industry have had the opportunity to testify at only one level, be it local, state, or federal. Most of the testimony has been by individual entities, specific trade associations, or affected elements of the building community. Now it is necessary to strengthen the way in which the building community itself can convey its opinion of proposed regulations in a more positive and stronger sense. There also is a need to develop a mechanism for disseminating information on emerging regulations at all levels and for providing the construction community with a voice in the regulation development process.

Some efforts are under way that could be duplicated in other areas. For example, in Los Angeles and Chicago the building community has a direct way of participating in the code-making process at the local level. Other cities also have done this. This is something that people in an area can bring about by joining together. In addition pre-application reviews and conferences can be held to eliminate delay when a plan actually is submitted for approval.

These are just some of the themes of the Panel's discussion. What will be important from this point on is determining how the building community itself, in all its various facets, can come together to ensure that the benefit of regulation is realized without an adverse impact on the building process.

SESSION IV: HUMAN MOTIVATION AND INCENTIVES FOR IMPROVING PRODUCTIVITY

MERLIN L. TAYLOR
Assistant to the President
International Union of Bricklayers and Allied Craftsmen
Washington, D.C.

The Panel had so many divergent views that I will try to summarize each speaker's remarks. The chairman, Thayne Robson, indicated that the standard of productivity against which most nations measure their own is still the American worker. Productivity, or the output of construction today, is significantly different from what it has been in the past. Construction products now must meet stringent health, safety, and environmental standards, and the building process must cope with the regulatory apparatus. New methods, technologies, and materials have been utilized successfully in the construction industry, but the application of new human resource management techniques in construction has lagged. Dr. Robson indicated that positive steps toward improving productivity can be taken in: (1) training (managerial instruction and skill upgrading for craftsmen as new techniques are brought in on the job), (2) improved scheduling, and (3) labor-management cooperation (through pre-job conferences, improved work facilities, project agreements, conflict resolution mechanisms, improved communications, and job review).

Mr. McArthur highlighted the scheduling problem as the biggest one faced by his firm, the Austin Company. The secondary consideration was manpower and, of course, the loss of tools. Training, he noted, is one of the most important aspects of a sound program. He pointed out that engineers must be retrained by the firms to do the job required. He also noted that 130 of the 150 superintendents of the Austin Company came from the trades or have been through apprenticeship and that getting good foremen is a major problem. Turnover and absenteeism also are problems, but the hiring hall arrangement with the various unions has provided some relief with respect to the latter. He rejected the notion that workers should be treated like pieces of equipment and turned away when not needed. Finally, he highlighted the need for pre-job conferences and mechanisms to eliminate work stoppages and handle accidents and job problems.

Mr. Kellstrom presented the subcontractor's viewpoint. He pointed out that wages and work negotiated in the contract should be adhered to by signatory subcontractors and that overtime should not be used as a raiding tactic. He also explained that neither the union nor the contractor should tolerate disruptive elements on the job. He noted that the contractor should make the very best tools and equipment available to craftsmen and that personal recognition improves morale. He also mentioned that supervision is most efficient when a foreman is a skilled craftsman who has been trained to handle the job.

Mr. Schmitt discussed the problem of government regulation and presented information concerning management and the utilization of incentives for an individual to perform a function properly at the hourly wage rate as opposed to a piece work rate. He pointed out that the alleged productivity decline is most likely a deterioration of the management function and that government has inhibited motivation and innovation. Divergent segments of the industry, he explained, have more in common than they did in the past and should join together to protect all aspects of the industry.

Mr. Georgine spoke about problems with current productivity statistics and their interpretation. With respect to planning, he stressed that craftsmen should be involved in pre-job and management planning of projects and utilized as a repository of skills and insights. He also talked about providing workers with simple incentives that might have a dramatic impact on job-site productivity.

Mr. Knapp touched on the fact that the Department of Labor, through the Bureau of Apprenticeship and Training, is deeply involved in developing proper standards for training. He also pointed out that the Department is responsible for assisting in the provision of a safe work environment. He commended the work of the Construction Coordinating Committee in dealing with the problems of seasonality and cyclicality through coordination of industry parties and in the letting of government contracts to exacerbate cyclicality.

Mr. Curtin indicated that strikes occurring during the term of the contract have a great impact on the legal aspect of labor-management relations. He cited statistics indicating that 57 percent of all work stoppages between 1962 and 1973 occurred during the course of the contract agreement. He discussed special project agreements in detail and identified four principal ingredients: a comprehensive no-strike clause that reflects a mutual promise between the two parties, a mechanism for enforcement, an all-inclusive grievance procedure, and a dispute resolution mechanism. He emphasized the benefits of expedited arbitration through which an arbiter can help a contractor to receive direct relief from the courts. He pointed out that this procedure is adopted on a voluntary basis and that its benefits are apparent in that only 23 percent of all strikes since 1973 have occurred mid-contract.

Professor Lewis very dramatically called attention to the need to recognize and deal with change. He spoke of public demand and public awareness and summed up his own educational experience in motivating community interest in the environment through participatory projects dealing with land-use planning. He emphasized that motivation to make collective work more productive arises from actual participation in the planning and in the execution of the finished work product.

SESSION V: INNOVATION FOR IMPROVED PRODUCTIVITY

JEFFREY HALLETT
Director, Productivity Center, Economic Division
Chamber of Commerce of the United States, Washington, D.C.

It was stated during the Panel's discussion and, in fact, many times during the Forum that three important things must occur if there is to be innovation in the construction industry:

1. A continuing willingness to share ideas about what works and about what other people are doing.
2. Technology transfer--finding ways to move technologies and procedures from one type of construction process into the rest of the industry.
3. The continuing evolution of a philosophy concerning what is being done so that attempts to construct projects in a useful and profitable way will be responsive to a changing environment.

It also was emphasized that to achieve meaningful productivity gains and generate the kind of innovations that have the maximum benefit, the final product must be defined more articulately. It must be understood that the value and the utility of a product of the construction industry extends far beyond the point at which the last worker leaves the site. It was also pointed out that a major survey of the industry revealed very clearly that there is serious concern for and interest in pursuing the productivity issue further, and examples of efforts of this sort were given.

One means identified to foster innovation would be an innovation in itself. In an industry characterized as being poorly capitalized and very fragmented, the motivation to invest in innovation, significant innovation, is retarded because the innovator gets only short-term benefits; new ideas tend to spread through the industry very quickly and there is no viable way for an individual to market those he develops. It was suggested that an institution to foster and stimulate such marketing and dissemination would be useful.

Communication and feedback among all the parties involved in the process were identified repeatedly as the most critical elements in stimulating continuing streams of innovative ideas and productivity improvements. It was noted that the feedback and communication process will benefit if attention is paid, in a quantitative way, to what is going on at the work sites and if some quantitative measures of performance are generated. The panelists made clear that the kinds of issues being considered at this Forum are not strange or unique to this country and that it would be a good idea to maintain closer dialogue and communication among us as we try to sort out ways to do better.

Productivity was defined as a state of mind and Japanese efforts were described, illustrating what happens when a whole society and economy gets the "right state of mind." It was pointed out that some of the benefits of a long-term commitment between and among organizations and their employees have to do with commitment to quality and commitment to the long-term health of the organization.

Finally, it was made clear that we are beginning to generate, through the American Productivity Center and elsewhere, the capacity, interest, and support needed to continue to generate information and assistance for improving productivity. Needed now is a greater demand for those kinds of services. The capacity to deliver them is in place.

Session I

MANAGING THE BUILDING PROCESS FOR IMPROVED PRODUCTIVITY

PANEL OVERVIEW

RICHARD P. GODWIN (Moderator)
Executive Vice President
Bechtel Incorporated, San Francisco, California

This Panel has been asked to address a subject that is stated in the Forum program in somewhat disarming simplicity--"Managing the Building Process for Improved Productivity." Having been a member of the planning sessions that preceded this Forum, I can attest to the fact that there probably are no two people here who will agree precisely on what the building process consists of or on what this particular title means. This is a reflection of the diversity, large size, and somewhat disjointed nature of the industry. Actuallly, however, given the complexities of the building process and its close relationship to the nation's economy, it is not surprising that we have a diversity of interests and a diversity of inputs. Accordingly, the Panel on Managing the Building Process for Improved Productivity has been assembled to represent a microcosm of the industry.

Regardless of its diversity and complexity, it is a big industry that, as was mentioned earlier, represents one-tenth of our gross national product. Figure 1 depicts the scope of the building environment and breaks down the industry into the following categories: residential, institutional, commercial, industrial, recreational, and infrastructure. Home construction or construction related to residential construction accounted for about 50 percent of the $220 billion spent in 1978. About $33 billion or 11 percent of the total is devoted to commercial construction; $17 billion or 8 percent to institutional construction; $28 billion or 14 percent to industrial and recreational construction; and $40 billion or 20 percent to the infrastructure. Of the $220 billion, $157 billion was spent by private sources and the remainder of 44 billion dollars was spent by government. In addition, the U.S. construction industry did between $20 and $25 billion worth of overseas construction in 1978. Finally, as others have mentioned, 1 out of every 15 persons in the labor force is associated, directly or indirectly, with the industry.

SCOPE OF THE BUILT ENVIRONMENT

(1978 CONSTRUCTION SPENDING - $202 BILLION)

RESIDENTIAL
- SINGLE FAMILY
- MULTIFAMILY LOW-RISE
- MULTIFAMILY HIGH-RISE

COMMERCIAL
- OFFICE
- MERCANTILE
- HOTEL/MOTEL
- SERVICE/REPAIR

INSTITUTIONAL
- GOVERNMENT
- HEALTH CARE
- EDUCATIONAL
- RELIGIOUS
- RESEARCH

INDUSTRIAL
- RESOURCES PROCESSING
- POWER GENERATION
- PRODUCT MANUFACTURE
- FOOD PROCESSING

RECREATIONAL
- SPORTS
- PARKS

INFRASTRUCTURE
- SANITARY
- CIVIL WORKS
- UTILITIES
- TRANSPORTATION

FIGURE 1 Scope Of The Built Environment
(1978 construction spending-$220 Billion)

Just considering the implications of an effort to apply a uniform management approach to this wide diversity of projects and types of work leads to the problem we will discuss here. Obviously, standards in both design and specification and for mass production would be easy answers for us if we could get to that point. It is somewhat curious that the building industry over the years has had very little success with mass production. Prefabricated housing has done poorly in the United States, and this persists around the world. In Saudi Arabia, for example, where construction is going on at a monstrous clip, prefabricated housing would be an obvious answer because of the lack of indigenous materials, but it is unpopular and somewhat less than 5 percent of Saudi residential construction is of the prefabricated type.

Not only is the building industry diversified in the sense that it is large and generates various kinds of products, it also has a large number of participants, each of whom views the industry from a different point of view. (Figure 2). Without elaborating too much on the variety of participants, let me just note that one's opinion about which is the principal activity of the building process tends to depend on one's role in the process.

PARTICIPANTS IN THE BUILDING PROCESS

OWNER-MANAGER

FINANCIER	PLANNER	MANUFACTURER
LAWYER	ARCHITECT	SUPPLIER
CONSULTANT	ENGINEER	TRANSPORTER
INSURER	DESIGNER	UTILITIES
AGENT	PURCHASER	PUBLIC SERVICES
	CONTRACTOR	
	LABOR-UNIONS	

REGULATOR
(FEDERAL/LOCAL)

FIGURE 2 Participants In The Building Process.

I will finish setting the stage for the panel discussion with Figure 3 which is a very broad definition or a generalized model of the building process itself. Although this model is obviously oversimplified, it does illustrate that every project must start with somebody's bright idea as to what he wants to build, what he needs, or what he wishes to construct over some period of time. This project concept goes through a planning stage and a preliminary design state, and, finally, a detailed design is constructed. Hopefully, it is completed and someone starts it up and runs it or occupies it. The broad groups of participants listed on the left consist of the owner, manager, and entrepreneur or developer; the engineer; the contractor and suppliers; and, finally, the owner again and the operator. Some people consider this a closed system, and I suppose in many ways it is. However, the point to make here is that the Panel will devote its attention to the upper left corner of Figure 3 and will discuss the contributions to productivity of the owner, manager, or developer and how they might be improved.

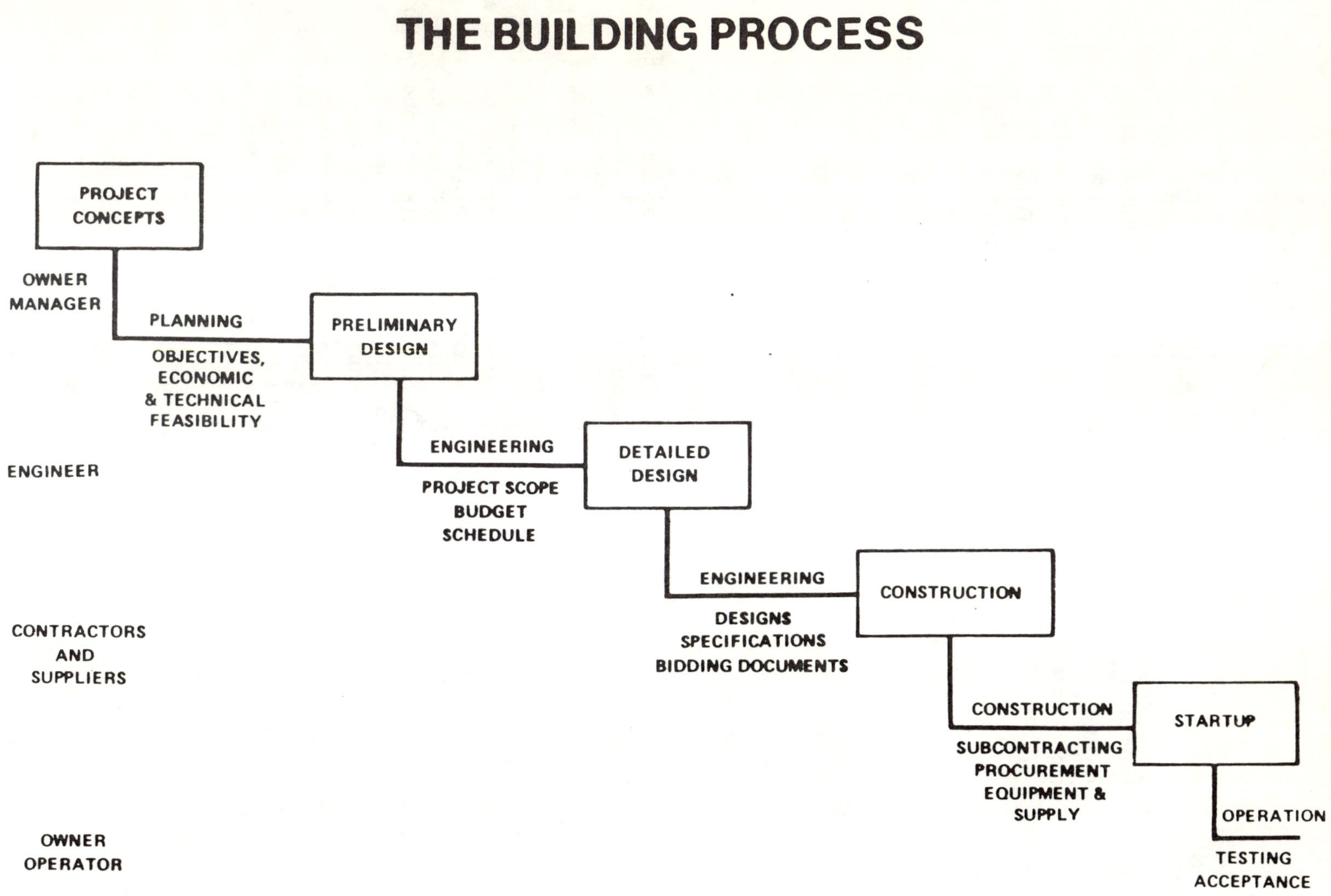

FIGURE 3 The Building Process.

Listed below each of the steps identified in Figure 3 are the deliverables. What do you expect to get from a project concept? Certainly the owner has to set the objectives and, hopefully, he assesses the technical and economic feasibility of what it is he's undertaking. These deliverables set the stage for development of a preliminary design, which more or less confirms the fact that what the builder has in mind is buildable and is financable. Detailed design specifications are generated, bidding occurs, and, finally, the owner has something that is tested and accepted. This sequence of events is followed no matter what is being built.

Figure 4 demonstrates that all of these steps are included in project management and is actually the organization chart for a project now going forward in Saudi Arabia, the first phase of which involves $27 billion and a management group of 1400 people. This should illustrate that all of the building process functions must be organized and planned by management if the job is going into be completed reasonably well.

The principal job of management is planning. Planning obviously consists of more than just putting pencil to paper. It involves determining the direction of the project, the rate at which it will proceed, and the goals; marshalling, controlling, and using of resources; and, perhaps most importantly, selecting the management who will carry out the job. Although the latter function may seem obvious, one must remember that the project will rise or fall depending on this selection. Thus, in the broadest sense, management's activities are to plan, to direct, and to control. The control is as important as the planning because a plan that is ignored is worthless. The planning, obviously, is most important in the early stages when most of the fundamental decisions are made and most of the problems are hatched.

The planning effort required at the outset of a project and the costs associated with it involve more than most people would want. Nevertheless, the costs associated with the initial planning and the marshalling of the resources are dwarfed in comparison with the difficulties and waste of resources that will follow if the planning effort is neglected. Most owners, developers, planners, builders, and constructors tend to underestimate the trade-off relationship between good planning and good results. Paper changes are, of course, the easiest kind to make; therefore, can change a project at the beginning quite easily. When you get to the concrete, however, a change becomes very expensive. In a cartoon I saw recently, a Cleopatra-type figure was talking to the master builder; in the background was an almost finished pyramid. Cleo was saying, "I just don't know. Let's see how it looks about 12 feet to the right." That is typical of the problem we have. If management's role is to guide, monitor, and control the progress of the building cycle, it had better avoid launching the project down wasteful and pointless paths and concentrate on developing the buildability and the practicality of the project itself.

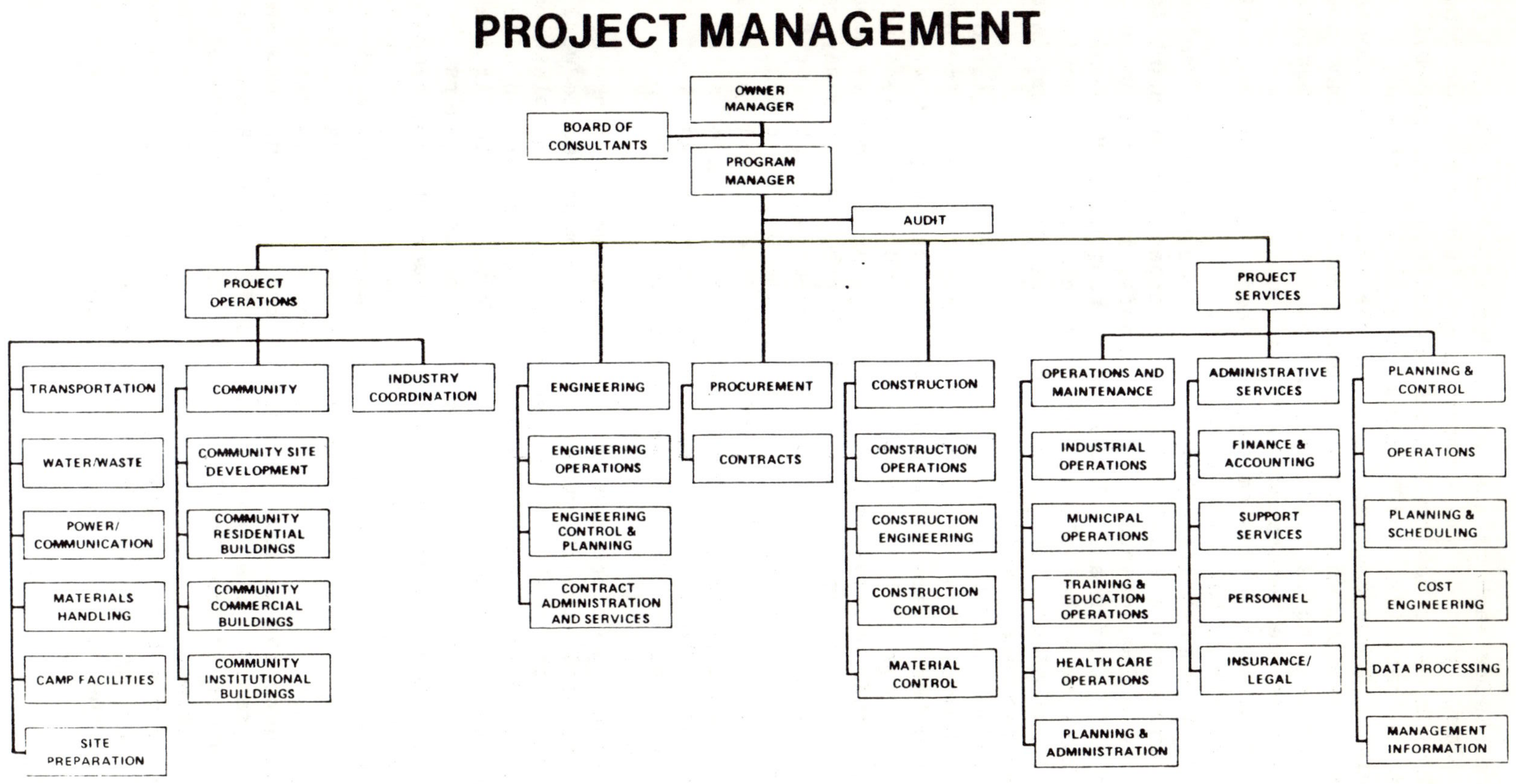

FIGURE 4 Project Management.

Among the major detriments to productivity in the building process are delays, disruptions, and deferrals. Some of us in the construction field operate on three commandments: First, unlike good wine, no construction project improves with age. Second, the principal theme should be "get in and get out." Once you have decided what you are going to do and where you are going to do it and, hopefully, have organized it correctly, the sooner you get in the field and the sooner you get out will be one of the principal reasons you come in within or even under budget. Third, the shorter the schedule, the more planning is necessary. On a man-hour basis this is certainly true. Another axiom probably should be stated: Putting a team together at the outset and having it available when you need it is the cornerstone of a successful operation.

If the principal role of management is in the planning process (i.e., in identifying conceptual or economic obstacles at the earliest stage), a master plan should be developed. This plan should apply to everyone, should be generally known by everyone, and should be in a form that will be understood by everyone on the project in progress. It should not be a Veda Bible; instead it should be a living and useful operation, something we normally identify as the "cookbook." It should be a simple, step-by-step, chronological arrangement of those things that have to be done to take a project from initiation to completion and placed in the hands of the people who will use it.

Finally, the building process has a built-in delay factor, regulations. This topic will be touched on later by other panels. I will note only that the size, complexity, and number of participants in the building process dictate that you "get the word out." Communication is important, but equally important is the need for management to recognize that the rules and regulations which govern the operation are not immutable and that we, as an industry, have been particularly lax in participating in the formation of these regulations. I think it better to meet the regulations in the formative stage, to participate actively, and to communicate with our legislators. We must develop a system within which we can at least forecast where we are going and identify some cost and productivity results we all can live with.

PRESENTATIONS

VANCE W. TORBERT, JR.
Architectural Officer
Metropolitan Life Insurance Company, New York, New York

I am going to talk about the building process as it relates to preplanning and program development. The success of any building project is founded on an efficient building program. Time spent in this development pays very handsome dividends, not only in the building process but also throughout the entire life cycle of the project. Unfortunately, heavy travail persists between the time the project is a gleam in the eye of the owner and when it is ready for occupancy and operation. Very often, the roots of these problems are based in oversights in proper program development.

There are many restraints to productivity, and I will mention only a few broad types. The basic building process is tremendously fragmented. It is really an overgrown cottage type of industry with very little coordination in any phase among the different groups. The basic elements are tremendously diverse and involve innumerable specialits, few of whom have a grasp of the whole process.

To compound the problem, communication between these various groups is, at best, very irregular. Of course, this is nothing new in the construction business. Remember the major biblical construction project, the Tower of Babel. You know what happened when all the specialists got together but did not speak a common language. That is pretty much what happens in at least part of the industry today.

The contributors to the program development process who should have relationships with one another and sometimes do to a minor degree are the owner, project developer, designer, construction manager, lender, and operator. The closer these organizations and groups work together, the better the final product becomes.

In the private sector, building owners (with the notable exception of large firms like AT&T and IBM that do a lot of building) are extremely ignorant and naive about the whole process. This is not bad in itself because there are people who can tell them which way to go; however, more often than not, we find that they do not know what their own requirements are. They know they want to build a building, but they do not know what the population of the building will be, whether they should anticipate future expansion, and things of that nature. They do not have their basic objectives well established at the outset. This can cause all sorts of delays later in the process because of concept changes and additions or subtractions. Thus, lack of a clear understanding of building function complicates the whole process and destroys efficiency.

There are two general varieties of owner in the private sector. First is the speculation builder who is just going to build a building to get in and get out. Heaven help the poor owner who eventually takes over its operation. Second, is the long-term owner who will remain, perhaps for the life of the building. These people have very different attitudes regarding building development, and what should and should not go into the building.

The owner should have an idea of the general character of the building--whether it is going to be a monumental type or just a minimum functional structure, such as an industrial processing building. Very often they are completely unaware of what the character of the building should be.

With regard to budget, owners have an idea of what they would like to spend but then are horrified at what they find the construction dollar will buy. Although they can get budgeting advice from experts, the budget is something that should be developed as input from them. One means owners do not utilize, perhaps because they are unaware of it, is the feasibility study. Such a study can eliminate considerable confusion and may even convey that a project is ill-conceived and should not be pursued. It can be tremendously valuable to an owner and can prevent eventual disaster.

Project developers also should provide program input since they can provide data on real estate conditions and site selection. They also should be versed in construction coordination and timing as it relates to economics, speed of building, and what is going to happen when the building hits the market.

The third category of contributor is the designer, the architect-engineer (A-E). The A-E's main function is to understand and interpret the owner's basic motivations for building and to translate them into the professional terms that establish the general scope and character of the project. The A-E also should be very cost-conscious (some are not) and should try to relate design criteria to budget. An important condition that is sometimes overlooked in the matter of economy is the fact that the owner, in order to save a few dollars, will not engage the A-E to provide the construction inspections that will ensure compliance with the drawings and specifications during construction and that will control costs. I hesitate to criticize my own profession, but A-E firms often lack personnel properly trained to fulfill this obligation. The A-E also should be aware of the lender's design and construction standards. This is something very often overlooked until the owner approaches the lender to obtain financing. Often the lender will have minimal standards of construction and design that are not rigid but that may change the scope of the whole program. Clearly, this knowledge ought to be introduced at the outset of the program.

The owner also must be advised of the proposed method of construction because it strongly influences project organization. If the fast track method is to be followed, specifications and drawings are prepared in a certain sequence. If conventional procedure is to be used, they are developed in another way. This fact should be brought out at the outset of programming, not down the line somewhere.

The construction manager is the constructor, and his input is extremely valuable. He will provide basic data on construction practices that should result in financial economies and time savings. His expertise should be utilized by the designer in the selection of materials and systems. More often than not, however, there is tremendous resistance to this practice on the part of the professionals. Obviously, the better the coordination between these groups, the better the end product. Here again, this should all be spelled out in the basic program.

Construction schedule and completion dates should be closely coordinated by designer, lender, and owner. They should be realistic to avoid such things as extra construction loan interest costs, occupants' moving expenses, and operational delays, all of which add to the final bill and decrease efficiency.

The construction and long-term lenders also are contributors to the process. Coordination between types of lenders is extremely important. It is a difficult criterion to express in an initial program but it should be considered. The construction lender, for instance, is anxious to get a quick take-out but also should be able to accept qualified approval of the scope of the project by the mortgagee. Very often the process is held up interminably by exorbitant demands on the part of the construction lender who wants to get out with a whole skin and have the long-term financier assume all the risk. As noted above, the design and construction requirements of the lender should be incorporated into basic design by the A-E.

Last, but not least, building operators should be, but rarely are, able to contribute to the basic concept of a building project. The operators, after all, are going to be stuck with the building after it is completed. Amazingly, we find that owners often have no real concern about what may happen to the project after the construction period. They presume that the building will just float along and become a great money-maker by itself. Not so. Maintenance and operation methods must be recognized in basic design. Will the maintenance be done by in-house personnel or contracted out? Under certain circumstances, space for maintenance materials and equipment must be provided in the building plan. Factors as mundane as window-washing method ought to be determined very early in the design phase for a high-rise building. The economic success of the whole project in the long term is tremendously influenced by advanced maintenance planning. This also ought to be written into the program but usually is not.

How can productivity be improved at the program phase? Each contributor to the process should be made aware of, and appreciate, the influence of the others on the final building product. The barriers and the prejudices of the specialists must be broken down not only by interprofessional dialogue and communication but also by cooperation. There is a dire need for a rebirth of the master builder who knows and can direct all elements of the building process from the initiation of the basic program and feasibility study to completion.

At the present time, there is a vacuum in this area. The gap is being filled more or less haphazardly by diverse categories of experts, developers, and construction managers with very mixed and, at best, uncertain results. One solution to this dilemma would be to develop an educational system to produce knowledgeable professionals who can coordinate all the divergent building forces into an efficient system. That is a formidable task. Metropolitan Life Insurance Company has endorsed this concept, however, by providing financial support for those who will work to obtain a graduate degree that prepares this new breed of designer-builder-financier. It has not been in force long enough to show results, but as far as we are concerned, this is the hope for the future.

J. L. ROSENBERGER
Construction Manager
Union Carbide Chemicals and Plastics, Houston, Texas

No doubt, we all agree that the management process is the planning and coordinating and the efficient marshalling of all available resources to produce a product. From the viewpoint of the owner-manager, the management process of the building industry must be considered as part of an overall goal of the company. It can be defined as the process by which a facility is completed at the location and within the timing and cost specified by the owner. As a result, the company's overall management approach has a strong effect on productivity in the building industry.

To show the relationship between owner-manager policies and building industry productivity, I will address three specific areas. Since my point of view is that of an owner-manager, the three areas relate specifically to an owner's organization; however, the underlying principles also apply to any organization involved in the building process. The three areas are:

1. The objectives and accountabilities of functional groups and project teams.
2. The establishment and use of a standard measurement and feedback system to provide a consistent and systematic method of communication and to generate credible progress and productivity data.
3. The continuity of chronological developments and historical data.

After describing these relationships, I will identify some action steps that owner-contractor organizations and the industry as a whole can take to improve productivity.

The first area affecting productivity involves the objectives and accountabilities of functional groups and project teams. Conflicts regularly occur as a result of differing objectives and accountabilities. These conflicts relate directly to the owner company's basic organizational structure and operational philosophies and procedures. Companies generally organize along functional lines (i.e., production, maintenance, engineering, materials procurement, distribution, accounting, sales, etc.). Each functional unit operates continuously, year after year, and the career path, supervision, and pay incentives of the unit's personnel are directly related to that functional accountability. On the other hand, the project team is composed of functional unit members and lasts only as long as a project lasts. The project team members therefore have dual roles: one as a team member and the other as a representative of their specific function. When the project team's responsibility ends, each team member returns to his functional unit and is measured by his loyalty to and interest in the unilateral concerns of that unit. If a company's operational philosophies and procedures do not take into account the dual role of the project team members, problems will arise. For example, the materials management function normally is held accountable for obtaining material as specified at a minimum price. The purchase of valves for a particular project therefore may be deferred until the total requirements of other projects equal the volume needed to obtain the lowest price. In the meantime, the project needs the valves to maintain the schedule in order to keep the construction crew working efficiently. The material team member representative may be forced to purchase some or all of the valves outside the system to meet the project's objectives. Doing so places him in a very uncomfortable position (i.e., by taking an action that may affect the performance of his functional supervisor, his own security could be affected). The net result is conflict and dissension that eventually will cause loss of productivity. Each company should evaluate its procedures and employee accountability systems and develop a program to ensure that operation procedures and individual accountability are related to both the functional and the project requirements.

The second area affecting productivity involves the owner company's system for establishing goals, measuring progress and productivity in relationship to the goal, and projecting the resources in time to complete the work. Again, part of the problem lies in the conflict between the objectives and accountability of the functional team and the project team. For example, the accounting unit demands feedback to fit its capitalization and product price costing objectives; the cost estimating unit demands feedback in a form to fit its estimating requirements; and the construction unit demands information structured to meet its labor and equipment planning requirements so that it can optimize resource utilization and stay on schedule. The tendency today is to satisfy the best organized and managed and most powerful functional unit, and, as a result, the team effort suffers. Team members who are not part of the strongest unit become defeated or develop separate systems of their own. Lack of communication and coordination results, causing the application of the wrong resource at the right time or the right resource at the wrong time. This has a severe impact on productivity.

Reconciliation of functional and project goals through the development of operational policies and procedures is part of the answer. A standard measurement and feedback system can further assist in solving the problem. This system should:

1. Provide work definitions that can be easily understood and that each team and functional unit member can relate directly to his responsibility and accountability.
2. Provide a consistent basis for estimating the work involved.
3. Provide a basis for identifying actual resources expended and relating them to the work accomplished.
4. Provide a basis for progress reporting that accurately reflects work accomplished.
5. Generate management information that clearly and precisely describes progress, productivity, and remaining resource commitment and that is structured to correspond to management and functional level requirements (e.g., a design function may report to the project manager in terms of concrete design, structural design, etc., but within each of the subgroups, such as structural design, there would be further work description breakdown and more detailed feedback systems to allow specific status reporting of that function.)

A standard measurement and feedback system is essential if the objectives of functional groups are to be made compatible with project goals and if a consistent and reliable means for supplying credible progress and productivity data is to be developed. The project team then can feel confident that its decisions are based on sound data. This obviously will help improve productivity through better more decisive management.

The third area that affects productivity in the building industry involves technological developments and historical data. In order for any system to work effectively and continuously, the need exists to provide a means for learning from the past, evaluating existing systems, and projecting into the future. In the building process, this can be done by reviewing completed projects, identifying the innovations used, and compiling the pertinent data for historical purposes. With this knowledge, current systems can be updated and communication, organization, and planning can be improved. The result will be continuous improvement in productivity.

However, with regard to the roles of the functional and project groups, the same problems arise in the technical development area. If the material unit is measured totally on the basis of its capability to obtain material, it has no incentive to learn from project experience. If there is no standard measurement system, the cost of one project cannot be compared with the cost of another because different things will be described. In this case again, the dominant function syndrome comes into the picture (i.e., accounting criteria for capitalization). To solve this problem, the functional, operational, and administrative company management must devise policies that establish procedures and employee accountabilities consistent with the objectives of both the project effort and the functional unit. In addition, a mechanism must be created for evaluating project technological developments and historical data to determine how each functional unit can contribute to the overall project. When a full-cycle program consistent with these goals exists, the company can expect continued growth of management expertise. The resulting improved communication, organization, and planning ultimately will create an atmosphere conducive to continuous improvement in productivity.

Thus, it is clear that the overall management process of the owner company is closely related to productivity in the building industry. In order for owner and contractor companies and the building industry as a whole to effect an improvement in productivity, the action steps described below should be part of the overall strategy.

Owners and contractors should take two primary steps. First, each organization should establish a position or functional unit responsible for:

1. Evaluating the existing organizational problems and procedures of the functional units involved in the building process and of the project management teams and develop improvements.

2. Establishing a standard measurement and feedback system and guiding its implementation.

3. Analyzing ongoing and completed projects to identify innovations and good management practices.

4. Communicating those practices to in-house management, acting as an advisor to ongoing project teams, and training management personnel at all levels concerned with the building process (only if this is done can management expertise be kept current and updated).

5. Attending industry training sessions, seminars, and similar events. Second, owners and contractors should support improvement programs and ensure that both project and functional group work is a joint effort organized to accomplish the company's goal.

Three industry actions will contribute to improved productivity. First, the need for a construction research and development organization should be assessed. If it is needed, it should be established and assigned the following minimum responsibilities:

1. Acting as a consolidation or clearing house for construction technological innovations in management and engineering.

2. Developing improvement programs, preparing implementation plans including training programs, and distributing the results to the industry as requested.

3. Obtaining input from the industry on specific problems, performing needed research in these problems, and publishing reports of research results. Second, the need for improving productivity should be emphasized continously by encouraging and supporting special conferences, such as this one, as well as annually held training sessions, seminars, and national meetings. Third, management courses,

specifically courses focusing on management problems in the building industry, should be included in college and university engineering programs.

ALBERT R. MARSCHALL
Commissioner, Public Buildings Service
General Services Administration, Washington, D.C.

My position with the Public Buildings Service (PBS) is fairly similar to the one I had in the Navy with respect to my public owner-manager responsibilities. At PBS, however, I have a quite different heterogeneous group to worry about. About a week before I retired from the Navy, my master chief petty officer of the Seabees, who was going to retire on the same day I was, asked, "Admiral, how do you feel about going out there and being a civilian?" I said, "Well, Johnny, I don't feel too bad. My mother and daddy were civilians and I guess they're all right." He said, "I'll tell you, Admiral. The thing that worries me is that there is nobody in charge out there." Quite truthfullly, I have discovered this to be the case. Trying to "get in charge," of course, is one of the problems of every manager. Since I am fairly new in my job and am still learning, my remarks will be a bit different from those of the other panelists.

The PBS operates about 10,000 buildings around the country, approximately 240 million square feet. The government owns a little better than half of this space and leases the remainder. One problem is that the leasing curve has been going up and the building curve, going down. We hope to rectify this situation because, if it continues, the rent bill will be $1 billion in 1985. Thus, the PBS is concerned about its building program. PBS's building objectives are the same as those of other public and private sector organizations--it wants to produce a quality environment within a building and its surroundings so that the people who work and visit there will feel pleased and productive.

Productivity in the United States has suffered recently, and I was quite interested in Dr. Siegel's comments about management as well as the labor force being to blame. Maybe management is not as directly connected with labor as it should be. In assuming my responsibilities at the PBS, I found that there are many ways in which I cannot affect the productivity of the work force; however, I can enhance the productivity of our dollar, and I guess that may be much the same thing. Mr. Gerstenberg of General Motors has said that productivity is a measure of management's efficiency in employing all the necessary resources--natural, human, and financial. He also said that the human factor contributes from 10 to 25 percent of productivity growth but he did not specify whether it was productivity in the work force or not.

Let me get back to what the PBS is doing. When I arrived, I found that there were a couple of things that slowed down the architect-engineer (A-E) selection process. One was the use of public advisory panels. Although these panels were composed of well-trained and responsible people from the private sector, their use caused certain delays in the system because we had to be guided by the schedules of these busy men. Therefore, one of the things that I have caused to happen is that we will no longer use public advisory panels.

The second thing that I discovered was that PBS was using a "level three approach" on major projects. That approach consists of narrowing down the field of A-Es to three, giving each of the three a feel for preparation of concepts, allowing them six months to prepare a project concept, and then evaluating the concepts and selecting a winner. This is a very time-consuming process, and at a time when construction costs are increasing at the rate they are now, we cannot afford the luxury of this type of selection process. Our management approach to this problem has been to go back to what we believe is the traditional government way of selecting A-E firms.

To get a good job from an A-E, one must be a good client, one must let the A-E know exactly what is wanted from him. It is incumbent upon the public owner to ensure that the A-E is told in great detail what is expected of him, and we had not been doing that. Last year we conducted a post-occupancy evaluation and discovered that no one knew what we had set out to do in the first place. There was nothing either in writing or in the corporate memory that indicated we had told the A-E what it was we wanted. We gave him a budget, and told him how many square feet of space we wanted, and sent him off to work. As owners and operators, we must ensure that we give the A-E the proper guidance and we must ensure, as we go along, that he is following this guidance. I think our productivity will go up considerably when this occurs.

In the construction area I discovered that the PBS had spent a great deal of time, money, effort, and publicity on the construction manager concept. I have no quarrel with the construction manager concept as practiced by some owners in some fields but I take great exception to the way it has been handled in the PBS where it seems to be a fifth wheel in the construction process. The necessary authority was not given to the construction manager, and the PBS did not demand from him the performance that it should. In the future, we will depart from that concept and go back to another time-honored and system-tried concept, that of the lump-sum competitive bid.

One of the problems with lump-sum competitive bidding is that there eventually will be a few change orders. One of the things that we hope to do is to restrict future projects to a maximum of four packages such as demolition, foundations, superstructure, and finishes. The finish package probably involves the most change orders. In the private sector, the finish work is part of the final contract because the private owner generally does not know who will occupy a building until it is almost complete. If a private owner generally does not know, you can well imagine that the PBS will never know (until it's too late) because of the frequent changes in government priorities. We hope that by splitting up the jobs, we will have more meaningful phases of construction and still eliminate the many problems with lump-sum bidding experienced in the past.

However, we have learned a lot about the total building process as a result of previous efforts. One of these efforts was building systems. I look upon that as an abject failure. Construction management as practiced by the PBS was not so good either. We do have some demonstration buildings that are pretty good. The one in Manchester, New Hampshire, has 11 different heating and cooling systems. It is an experimental bed, a platform, for us, and it is working out in just that manner. We are deriving a great deal of information from it. In Saginaw, Michigan, we have built a facility that is essentially underground. It has the largest surface solar collector of any of our buildings and is working out very well.

One of the things that we are doing right now eliminates the requirement for excess space and, therefore, affects productivity. We have come up with a systems furniture approach that has reduced the amount of space required for the individual worker by at least 20 feet and that, in the process, has made workplaces considerably more attractive and more desirable. A wing in the main General Services Administration building that was designed using this systems furniture approach was opened about a month ago. The people occupying it are ecstatic about it and seem to be working better. That, I think, is all part of the productivity package.

What we in the public sector have to do now is to learn from the private sector and to pass on to the private sector what we have learned. In the last analysis, we will not work very much harder, just a lot smarter.

ROBERT L. WILSON
President
Robert L. Wilson Associates, Inc., Architects/Planners,
Stamford, Connecticut

The building industry, as it has been described at this Forum, really does not exist. We have characterized it as the largest single industry in America since it represents 10 percent of the gross national product. Unfortunately, however, it is not an industry at all in the sense that the automobile or oil producers are industries. It is a fragmented, fractured compilation of entrepreneurs, mostly small, who together comprise what is called an industry. Its effectiveness is limited by the number of entrepreneurs involved at various stages of the development process. This system encourages inefficiency, waste, additional time, and excessive cost and results in a product whose quality continues to diminish.

As presently formulated, the building industry is an anomaly, a $250 billion economic stepchild and political football that is consistently used to fine-tune the economy because of government's misguided fiscal and monetary policies. The industry has no real political clout because it does not speak with one voice--a necessity if it is to be treated (read coddled) politically like Lockheed or Chrysler.

Another serious problem with our industry, as well as our entire economic system, is the regressive confiscatory tax structure that discourages capital formation and encourages, through inflation, a consumer-oriented rather than a production-oriented society. I believe that three steps can be taken by the federal government to help stem inflation and turn us again toward a production-oriented society: (1) reduce taxes by at least 50 percent, (2) eliminate government deficit spending, and (3) return the dollar to the gold standard.

Management, by definition, implies control of a process, the ability to steer a predictable course of action over a controlled period of time. The building industry at present is not really managed. Its cyclical nature, volatility, and continued dislocation all have a negative impact on productivity. Moreover, it is not possible to effectively manage the building process in its present structure. If we are to be an industry, it will be necessary to structure ourselves like an industry. It is not necessary to look very far for examples. The aforementioned automobile and oil industries may be examined for clues. Let us create some real giants in this industry--a General Motors, Standard Oil, IBM, and AT&T of the construction industry.

The structure I envision would be a diversified entity that would have the ability to plan, finance, construct, market, and manage products of the built environment. Insurance companies, lending institutions, and real estate, contracting, engineering and design, management and marketing firms all would be included. They could be individual profit centers, either wholly or partially owned subsidiaries of a conglomerate. The idea is to aggregate all of the responsibility under a single point with the ability for internal capitalization and financing.

An example of the possible markets is the housing industry. Housing constitutes 50 percent of the building industry and could be used as a good model. There is a predicted need for at least 2 million housing units per year during the 1980s to replace obsolete stock and service new household formations. Predictions for housing starts for 1980 are as low as 1.2 to 1.3 million, and the real demand obviously is not being met. An industry that could build, and finance, units on its own would have a market that guarantees absorption of its product. Strong market demand would help eliminate cyclicality, effect predictability, and, hence, encourage control of supplies, production and costs, the very ingredients needed for management control. Increased profitability, productivity, and quality control would be just some of the by-products of this system.

Obviously it is not possible to create a General Motors or an AT&T of the construction industry overnight. This is a long-range goal, but, in order to capture the market, to have an impact on productivity, a major step needs to be made in that direction.

Until we can do that, some other things to improve management control can be done. However, it is not my intention that such things should be done by the government. I do not believe we need any new legislation to do what can be done by the private sector. All we need are some people with the vision to see an opportunity and the ability to take advantage of that opportunity. Our moderator works for a major constructor and does that on a macro scale, but we need more of it, a lot more, if this industry is to have the kind of strength that is necessary.

Another possibility for increased management control is formation of a development team, a loose aggregation, as it were, consisting of the owner-developer, planners, architects and engineers, contractors, legal and financial planners, lending institutions, and a management team. This development team would band together to provide single-point responsibility for an identified market segment of the industry. Market aggregation by the team would give an ongoing economy of scale with growth potential on the local, regional, national, and international levels.

Because such an approach may not always be feasible, I have some suggestions for small businesses, those that make up the bulk of the building industry. Planning and management are vital to financial stability and growth. Proper analysis of a project prior to the commitment of large capital resources is fundamental to proper management. Scheduling and budgeting of a project are required on a constant and ongoing basis. Past cost records should be kept and updated with present experience.

My firm serves as consultant to a number of clients, including lending institutions. The point has been made that some designers, some architects, and others should understand the requirements of the lending institutions. As consultants to both lending institutions and developers, we walk on both sides of that street. In working with some developers and small producers of the built environment, we have found an absolute lack of management and planning. In fact, we now are acting as a consultant to a bank on a Connecticut condominium project worth more than $1 million. The builder literally "planned" the project on a brown piece of paper that he carried around in his back pocket. He had absolutely no schedule for the purchase of materials or for construction. We were called in to help him, and we had to devise specifications for the project, cash flow projections, and schedules for construction. The project is now on stream. Thus, planning is extremely important no matter how small you are. In fact, the smaller the job, the more important the planning. People seem to assume that you do not need good planning because it's only a $500,000 job, and that is unfortunate. Many people fail because they do not plan well.

If analysis and planning are done correctly, production control is aided. However, it remains necessary to monitor the work forces (i.e., those of subcontractors) on a regular basis. Post-construction review of a project is extremely helpful for future planning, but I know very few owners, developers, or builders who do that. Good recordkeeping also is essential, but most developers and builders hate to keep records of any kind. If you do not have all the skills necessary to accomplish these tasks, hire a consultant or a staff that does. It will be money well spent. Its implications on productivity, I hope, are obvious.

WILLIAM H. BAHRKE
President
Dravo Utility Constructors, Inc., New York, New York

The primary efforts of the firm that I am associated with are related to heavy industrial construction projects, such as power generating plants, that often encompass development programs of 10 or more years; therefore, productivity and its enhancement are of vital importance to our operation and our life. The remarks I will make are associated with productivity during the construction phase of a heavy industrial project.

To determine productivity on a construction site is a most difficult task. However, some studies have been made recently in an effort to develop a breakdown of labor activities and to determine the amount of time actually spent on direct labor during a normal eight-hour shift. In this case, the direct labor would be the time actually spent pulling on a wrench or plying metal during the process of welding.

Most of these studies have indicated the following breakdown of labor activities:

Activity	Percent of Shift
Direct work	32
Waiting time	29
Late starts and early quits	6
Tools and material transport	7
Travel	13
Personal breaks	5
Instructions	8

These percentages indicate a project has a composite staff of construction labor that contributes only three out of eight hours on direct work. The need to improve the productive time, the amount of time expended on direct work, is obvious. The cost and schedule benefits are easily identified.

The greatest amount of nonproductive time is spent waiting. This is 29 percent or the equivalent of 2-1/2 hours of an 8-hour shift. I suggest that the proper application of sound and construction management plans, adequate and well qualified supervision, and timely and sufficient engineering drawings and decisions can reduce waiting time to at least 20 percent and probably much less.

To illustrate how this would be reflected in dollars saved, let us consider a 1 million man-hour project with 19 percent waiting time instead of 29 percent. This would result in a productivity improvement of 10 percent or a savings of a 100,000 man-hours. At a composite craft rate of $20 per hour, a savings of $2 million in direct labor costs alone would result. Additional savings would result because 1 month in every 12 months of scheduled work would be saved. This means that equipment, tools, and facilities would be used for a shorter period of time and that the constructed project would be ready for use sooner, which would result in additional revenue and possible tax savings. All of this would be stimulated by a 10 reduction in waiting time. Similar benefits certainly would result from improvements concerning the other items of nonproductive time.

The building process historically has been considered to be only the period of actual construction but it actually is of much longer duration. In fact, the building process begins with project conception and continues through completion and utilization of the facility. Although the constructor may be from one to three years away from actual on-site construction during the conceptual stage, the building process really has been initiated.

The design phase should address the methodology of erection, material procurement, shipment and installation, and equipment and tool utilization for most effective and economical installation. Productivity concerns should be given considerable attention by including the construction management entity as an integral member of the design team.

When I refer to the construction management entity, I really mean the concept of construction management. This service can be performed by a professional construction management company, which is the best approach, or by the owner, the architect-engineer, or the constructor. The essential thing is that the functions of construction management be introduced very early in the project and be performed throughout. Construction management has been utilized on specific projects in the recent past but only on a limited basis. Now it is necessary for the construction management aspect to become more detailed and allocated and more significantly voiced during the early decision-making stage of a project.

Even when the construction manager has been brought into the picture early, his role was only to provide a constructability review of the project's design, which would address ease of construction, details to be used, schedule considerations, construction equipment and its utilization, physical layout, and eventual interface between individual contractors. If we wish to improve productivity, this limited preplanning role is not sufficient. The following aspects of construction management should be given attention during the concept and design phases of a project:

1. Labor availability and quality;
2. Competitive projects within the area and their impact on labor requirements;
3. Training programs including the appropriateness of the skills taught and the number of trainees to be made available during the time of the project;
4. Work schedules including number of days per week, hours per day, shifts per day, and overtime;
5. Crew sites and access and egress for people, equipment, and material;
6. Facilities for people assigned to the project;
7. Flow of material from point of shipment to final installed position;
8. Availability of qualified supervisors and the ratio of supervisors to laborers; and
9. Value engineering concepts applicable to construction items of work.

This is by no means a complete list of what a construction manager should be considering during the concept and design phase of a project, but it does illustrate the degree of construction manager involvement desirable during the very early stages of a project.

This planning should be incorporated in the overall project plan, which should result in a realistic construction effort and realistic staffing requirements. The necessity for detailed planning cannot be emphasized enough. Without this type of planning, the completion of a project on schedule and within budget is sheer luck.

Subsequent to this planning effort and prior to any on-site activity, a concerted effort must be made to secure a meaningful project labor agreement. This agreement should be negotiated to best address the requirements of the specific project labor market. Significant aspects of the agreement should address labor productivity. To date, most labor negotiations tend to be focused almost exclusively on financial considerations (wages, fringes, and premiums). We now have reached a crossroads where productivity, or the lack thereof, is a primary cost factor, and a project labor agreement should emphasize such aspects of productivity as foreman selection, elimination of restrictive work practices, elimination of featherbedding practices, uniform shift work provisions, uniform holiday provisions, and elimination of jurisdictional work stoppages. This is just a partial list of items affecting productivity that can have a more drastic effect on job cost than normal negotiated wage increases.

We must attempt to create an atmosphere in which labor and management can work together to increase productivity. In a recent speech before the American Productivity Center, Robert A. Georgine, president of the AFL/CIO Building and Construction Trades Department, stated:

> A productivity program which has been hammered out in the give and take of bargaining is also more likely to be successful. The negotiation process itself may lead to improvements to management's plans as its use of labor are made known. Certainly employees are much more likely to cooperate enthusiastically in a program which they have had a real voice in establishing.... I firmly believe that labor is ready, willing, and able to continue to do its part in the future in meeting the challenge of productivity.

Thus, it behooves us to actively pursue a project labor agreement early in the building process. During the negotiations we should insist on union involvement in job productivity. This involvement should be at the local, area, and international levels. If Mr. Georgine is correct in his statement that labor is ready, willing, and able to do its part, the combined effort of both labor and management will be a step in the right direction.

During the construction management planning phases, managers and owners can create an atmosphere that is conducive to improving productivity. This atmosphere can be generated by providing adequate and clean sanitary facilities, cool and decent-tasting drinking water, a clean and dry place to change clothes and eat lunch that is in close proximity to the actual place of work, adequate parking with service roads leading to and from the parking area, and mechanical means of transporting the workmen to the place of work when the building is over five stories high. These items might sound unimportant and insignificant, but the lack of any one of them can cause a morale problem that will result in a large cost overrun.

The importance of a knowledgeable and motivated construction supervision staff must not be minimized. The time to designate the key members of the construction team is during the concept and design phase of the project. Supervisors who clearly understand the details of the project, the objective of the overall effort, the labor market, the project's schedule, and the project's budget will enhance the eventual success of the project. Supervisors who are high on the learning curve prior to the start of field work will be able to concentrate on craft labor productivity.

It is important that much of the project engineering be completed prior to the start of construction. Changes and revisions made to drawings after construction has started and the affected material already has been installed have an extremely negative effect on productivity not only because of the man-hours and time required to make the change, but also because of the detrimental effect on the attitudes and morale of the work force.

Computer programs have been developed that can greatly assist in the design of a project; available are computer-aided design and engineering (CADAE) programs, cable management programs, and computer-aided design and manufacturing (CADAM) programs. These programs and the many others I could mention have been developed both to reduce the number of man-hours required to engineer a job and to reduce the possibility of errors and interferences.

Many management tools for planning, scheduling, and cost control also are available (e.g., Project SCOPE and major milestone schedules, Project Master Control networks, work breakdown structures.) A project schedule is a dynamic working tool, not a static wall ornament. At Gibbs and Hill and Dravo, we print out a detailed two-week look ahead schedule every week, and job meetings are held with craft superintendents to review the items of work to be performed during the two-week period and the manpower and tools required. Progress from the previous week is reviewed, the schedules are reviewed, and procedures are established for catching up, if necessary. The important point is that these management tools are used to keep the field changes to an absolute minimum and to plan and schedule manpower, tools, equipment, and material to obtain maximum benefit from each.

During the on-site construction phase, the advance planning must be implemented in a manner that will establish productivity tone and motivation and maintain them throughout the project duration. To maintain this tone and motivation, we should employ the labor force using the most modest and gradual increases without jeopardizing the schedule and force production standards and project work rules. Emphasize the selection of foremen. Plan tools and equipment utilization, material flow, and manpower assignments to minimize waiting time, traveling time, and other nonproductive activities.

Good sound planning and execution is an absolute necessity if a project is to be completed successfully. A good rule to follow is to plan your work and to work your plan.

DISCUSSION

QUESTION: Mr. Rosenberger noted that owner-user organizations generally focus on major operating functions and that project management is therefore somewhat neglected. Adm. Marschall, on the other hand, indicated that the Public Buildings Service (PBS) is dispensing with the construction management concept. However, does not construction management help to alleviate a problem for those firms that do not have a large enough construction program to support an in-house construction management staff?

ADM. MARSCHALL: What I actually said was that construction management as practiced by the PBS was not very good. When I worked in the private sector as a vice president in charge of construction management, the concept worked well. It is the way the bureaucracy developed construction management procedures within PBS that I do not support. In addition, the PBS does, in fact, have a professional staff that can monitor construction and administer contracts awarded to the private sector, and it was the conflict between this professional staff and the construction management firm that made construction management in the PBS ineffective. We do not intend to disavow construction management completely--just the way it has been practiced recently.

MR. ROSENBERGER: In the past, the owner tended to delegate responsibility--to assign the responsibility for his capital expansion programs to an outside management firm, to a construction management group, or to the building constructor. More recently, many owners have come to the same conclusion as Adm. Marschall. They have decided that continuity, as far as the building process is concerned, rests within their own organizations; that they must create an organization to manage the building process; and that the continuity of building regularly will strengthen the expertise within their organizations and eventually result in better productivity. They have realized that they lose continuity if they use different construction managers for different projects and that their own personnel do not have the opportunity to develop expertise.

MR. TORBERT: This is all well and good for firms like Union Carbide and IBM that do a lot of building and can use in-house personnel and support them on an ongoing program. In the private sector, however, many projects are one-shot deals. What is the poor owner supposed to do in such a case? He has no choice but to farm out the job and to get the expertise he needs from the supposed professionals. That is where the owner lacks capability; he needs the fellow who can put everything together and create a viable product.

QUESTION: The importance of feasibility studies as part of the planning process has been mentioned. What are the ingredients of a feasibility study? Does it consist mainly of technical and economic factors?

MR. TORBERT: The components of a feasibility study naturally vary greatly depending on the type of project. Generally, the aim of the study is to identify the economic problems, social problems, real estate problems, and, sometimes, even certain design elements that will be fundamental to the project. It is a tool for a non-knowledgeable owner in that it results in a compilation of very basic criteria for the eventual production of the building. It does not define exactly what the design should be but rather establishes the basic parameters of the project.

MR. WILSON: From my standpoint, the feasibility study permits me to determine whether a project is a "go" or "no-go" situation and, if it is "go", to identify some of the problems that might be expected. We often find that an owner needs a feasibility study but does not realize it because he does not know what his problems are. We identify those problems and often find that, in stating them, we have found the genesis for a solution. There are times when a nonfeasible project becomes feasible simply because the problems can be identified and analyzed and tentative solutions can be established. The feasibility study ethic is one that I certainly support, and I believe it should be used no matter how small a project is because it is such a good planning tool.

QUESTION: A Japanese construction firm was low bidder for a major project in San Francisco, California, and will be subject to the same working conditions as the U.S. contractors. Will the firm actually be able to manage to achieve greater productivity or did some other factor give them the edge in the bidding?

MR. BAHRKE: I am familiar with the project and understand that the firm presently is experiencing financial problems on the job. These problems probably have resulted because the firm is using the same type of labor that a U.S. firm would and is experiencing some of the same management problems. Thus, its success in getting the job resulted from a good estimate. Its success in completing the job is still in question.

MR. GODWIN: Since San Francisco is my home, I know something about the project, and I believe this is the first time that a major Japanese contractor has bid on heavy civil construction in the United States. The project, a large storm drain system, probably involves a $1 billion construction program and is a major disruption in the city. The Japanese bid it as they normally do, as a group of companies headed by one, against U.S. construction companies in San Francisco. Although I am sure the Japanese did some head-in planning, the inputs they used probably were derived from work in Japan or other parts of the world, not in the United States, and the problems they are experiencing are those we know so well--regulation and labor. They did not import labor, other than certain key supervisory personnel, and, consequently, they are suffering some major delays because of regulations that they were not aware of and the disruption to the city that they had not anticipated. It will be interesting to see how they manage. I am told that other Japanese companies are interested in doing work in the United States, as are the Koreans, but now these groups are looking for U.S. associates to bid with them.

QUESTION: Adm. Marschall, how do you view the open office planning concept for government buildings? Is it more productive?

ADM. MARSCHALL: It is still in the test period, but the people who have been exposed to it seem to be very pleased with it. Personally, I like to be able to close my door, but I am 68 years old. I was astounded when those occupying the first open plan wing in the GSA building began talking about what a wonderful setup they had. They feel that they have privacy; they feel they have airy surroundings that are extremely nice. Open office planning does decrease the number of square feet required for each person, and, in that sense, it should cut our costs in the future but that has not been fully proven yet. The cost of a space with the new systems furniture is about $2250 per person whereas a normal arrangement is estimated to cost $750 per person. However, if our goal is to save space, we have succeeded. It may be that this saving of space causes the work force to rebel someday, but right now it looks good.

QUESTION: Will the integrated ceiling and background masking sound subsystem concept be continued in government buildings?

ADM. MARSCHALL: We really leave that decision to our clients. In the test bed at the GSA building, the users chose not to have the noise. When I was first exposed to the system in Toronto, I was amazed; in a work space you were not bothered at all by conversation in normal tones. There was a lot of wall covering, carpeting, and other things that helped, but the piped-in noise really seemed to make the difference even though I was not aware of hearing it.

QUESTION: Mr. Wilson, you mentioned the formation of major large engineering-construction companies to compete more effectively in the United States and abroad. We now are competing internationally with groups of companies from Japan, Germany, and France that have government participation. Is this what you have in mind?

MR. WILSON: No, I was referring to government participation only in the sense of nonconfiscatory taxation situations. We probably can be competitive technologically and from a management standpoint without direct governmental intervention or handouts. What we need the government to do is to return to more of a free economy with a less onerous taxation situation. I believe that would start to give us some of the advantages of productivity and the ability to compete in an international market.

QUESTION: Given the Forum's objective of developing doable recommendations, do you believe your political and economic proposals meet this criterion?

MR. WILSON: I threw in the economic and the political because I find it impossible to address the problem of productivity in the construction industry without taking into account the very wild swings and cycles in our economy that are caused by the government's fiscal and monetary management policies. We now have a 15-1/4 percent prime interest rate, and a number of projects are being financed at three points over prime. The numbers do not even work out at 15 percent let alone at 18-1/4 percent. This situation has resulted from a political decision. Some of the measures I suggested reflect my personal beliefs, but I think that my beliefs are just as valid as those of Alan Greenspan or Paul Volcker at this stage of the game because those gentlemen obviously have not been able to control the economy, to control inflation. Bechtel exists. Bechtel is not a governmental agency. If the government can establish such corporations as the National Housing Partnership to go into private industry and, in this case, supply housing, why can't I, as an individual who has identified a need for 2 million housing units, get together with Metropolitan Life Insurance to do the same. The insurance companies are the ones with money, the ones providing long-term mortgages. They sell an almost intangible service--they aggregate money. They often ask for participation, so why can we not formalize such a relationship. The government does not have to do it; in fact, the government should keep out of it. That is basically what I want--less governmental intervention and involvement and a lot more of private enterprise doing what it can do best.

MR. TORBERT: But what happens whenever we try to get any big groups together? We face a government charge of monopoly. This has happened even when we tried to organize some basis for developing construction standards that would elevate the industry. We could not do it. The same principle would apply if, for example, the 10 biggest construction companies decided to get together to enjoy efficiencies of buying and administration. The government would consider it a monopoly situation and order us to stop.

MR. WILSON: I suggest that many things do not work because arbitrary political decisions are made. Much of what the Department of Justice does is arbitrary, capricious, and frivolous and has no place in the reality of the marketplace. But political decisions have been made. If, for example, the Justice Department goes after IBM, it can fight IBM forever and not win because IBM has enough clout, enough backing, and enough money to fight the government to a standstill. The construction industry does not have that kind of clout because it has not formulated that single voice to talk for it. I spent two years on the National Construction Industry Council and most of that time was spent squabbling. We very rarely had ideas that we could all support. Labor, for instance, was totally excluded from that group so we obviously started off crippled. We have not found a reason to have the single voice we need to exercise the political clout that would, in point of fact, get the Justice Department off our backs. But let us try. If we cannot do it at that macro scale, let us start at a micro scale and keep building up.

QUESTION: Adm. Marschall, why do you consider the building systems approach a failure?

ADM. MARSCHALL: Money. The Navy tried it and the General Services Administration (GSA) tried it. Both found that it is more costly than conventional methods of building. These building systems have been experimental; we learned a lot from them, and that was our intent. You really have to be involved in big projects to make the systems approach work.

QUESTION: Why does the government continue with a 10 percent retention on buildings?

ADM. MARSCHALL: It does not. According to regulations it is 5 percent for the GSA. The Navy uses 10 percent until the job is half done and then reduces it to 5 percent. I strongly believe that money is the life blood of the industry. If we are to encourage good builders to do work for us, we have to ensure that the money flow is there. Money costs money. I have no intention to adhering to a policy that strangles the industry, particularly now.

QUESTION: What does the GSA consider to be an acceptable percentage of unused assignable space?

ADM. MARSCHALL: That has not been a problem recently. I am told that we now have a deficit of 12.5 million square feet of space in the agencies of the federal government that we serve. I do not believe it. My experience leads me to believe that the real deficit would be only two thirds or three fourths of the figure cited. Nevertheless, we definitely are in a deficit position. I think that government would have to shrink before we would have vacant space not assigned, and I doubt that my children, grandchildren, or great grandchildren will see that. The number of government workers in Washington, D.C., has increased by an average of 3150 annually since 1959 despite the promise of any number of administrations to cut government.

Session II

FINANCIAL PLANNING FOR IMPROVING PRODUCTIVITY

PANEL OVERVIEW

ROBERT C. HOLLAND (Moderator)
President
Committee for Economic Development, Washington, D.C.

A good many people today are very interested in what will be said here. Any chance that this might be thought of as a boring and uninteresting topic was taken care of by the Federal Reserve Board (FRB) about three months ago. Once the FRB started moving up interest rates and squeezing down on whatever it decided was the money supply, it moved financial planning into the heart of the current concerns of many people, not only in the construction industry but in other areas as well.

Indeed, the uncertainties the FRB has generated and some of the trepidations it has hatched have stimulated a long list of interesting remarks about tight money and what it is. One more was added to that collection earlier when Dr. Siegel referred to the economy as "taking hemlock these days."

If what tight money is doing to the economy is serving hemlock, this is the "hemlock cocktail hour," and our bartender is that tall fellow at the FRB who keeps saying that money is going to get tight enough to slow down activity. When people ask him how high that means interest rates will go, he generally says something to the effect that increases will continue until some people change their minds about whether or not it is worthwhile to borrow. Such statements cause problems for financial flows in many places; however, the FRB seems to maintain that such problems are not as bad as the ones we would have had to live with over time if it had not acted.

Money and productivity may not seem to be synonymous or even related terms. Productivity usually tends to be thought of in terms of physical things--brick, mortar, man-hours. How do you make money more productive? Get more of it? Buy more of it? Pay less for it? The Panel on Financial Planning for Improving Productivity will focus on how to use money realistically to improve the physical productivity of the building industry. We will discuss money as a kind of a paper lever that, within limits, can add to the power and to the productive use of the resources involved in building all types of structure. We also may accentuate the negative occasionally by talking about money in terms of how to handle it so it least detracts from productivity when you cannot get it. Let me begin, however, by identifying the financial issues and the financial obstacles to greater productivity that caught the attention of those of your colleagues who were involved in the Forum planning workshops.

Major emphasis in these workshops was placed on recognition that building industry financial problems come in two sizes and that it is important to distinguish between them. First are giant economy size problems--financial problems at the national level that are related to the overall availability of investment funds and the portion of those funds that is used to finance the building industry. Second are company size problems--financial problems at the individual firm level that are related to how a firm can best arrange for a stream of financing for itself and its customers, how it can plan for changes in that stream, and how it can use that stream of funds most efficiently.

It is not hard to pinpoint why the building industry is plagued with giant economy size problems. Construction is, as we have been reminded, the biggest single user of savings, especially long-term savings, in this country; therefore, any development that significantly changes either the total flow of savings or how that flow is divided up can have a great effect on the amount of funds available to the building industry and individual builders. If anything adverse happens, mortgage money becomes tight, interest rates increase, and your friendly banker becomes a little less friendly. Thus, the building industry has a great stake in policies that can make the total flow of savings larger and more dependable because its own construction plans can be thrown very badly awry by forces that discourage or disrupt this flow.

It does not take much imagination to identify some of the key forces that disrupt the savings flow. Earlier we heard about tax policies, especially those that bite harder on saving than they do on consumption. We have built up a gradually increasing consumption bias in the tax structure of this country and in many other aspects of our economic system as well, and that bias is tough on home building. It is tough on building of all types.

Another troublesome force is inflation. If inflation does not continue for too long and if it is not too fast, some people in the business like it. It can seem to encourage saving. It can seem to create a gently rising tide of real estate values that will put smiles on the faces of a lot of real estate developers. But it is hard to keep inflation that benign. When it becomes rapid over a long period, it can make saving money look like a fruitless exercise. It can cause interest rates to increase rapidly. It can create a "boom and bust" potential inside the industry that is very painful indeed. Most of us can recall episodes--the 1920s or 1974-75--when the entire construction industry got squeezed very hard, both during the peak of inflation and during the subsequent downturn of the nasty inflation-recession rollercoaster.

The building industry can suffer a kind of "double whammy," to use Al Capp's phrase, when inflation is being driven by some other especially strong competitor for savings. For example, if the government is running a major deficit at a time when we are riding through this kind of rising tide of interest rates, its deficit financing will pull money away from other savings users such as the building industry. Much the same can happen if major corporate business is on a borrowing binge; a good many otherwise good construction credits can find themselves shouldered aside. In such circumstances, the money squeeze becomes doubly hard.

It is important to think carefully about how to solve these problems. Too often we have proceeded as follows: We have a problem. We ruminate on it. We want to do something about it. We introduce some measures that we think are going to solve the problem. Then we discover that these measures produce unintended or unappreciated side effects that rapidly escalate into full-scale problems in their own right. The building industry has had many experiences of this sort because it has been effective in obtaining a number of major changes in public policies directed toward construction.

For example, to enlarge and protect the amount of national savings going into housing, we passed a series of laws encouraging specialized financing intermediaries, like savings and loan associations, whose lending was essentially confined to home mortgages and whose financial health was protected by a series of regulations covering the terms and conditions under which they could obtain and lend money. That approach worked well in many respects; however, during the past 15 years big cyclical swings in interest rates, around a rising trend, have squeezed these institutions harder and harder against their regulated interest rate ceilings. That squeeze process has dried up mortgage funds repeatedly and put the brakes on building activity. To solve this problem, we have spent a good deal of time during the 1970s loosening interest rate ceilings, building secondary mortgage market makers, and creating adjustments that would ease some of the legal and regulatory constraints on the savings and loan industry and the related savings institutions. What has happened as a result? As we eased constraints during the latest upsurge in credit demand, we found interest rates going even higher. As we reduced the kind of rationing and allocation effects that were built into the earlier set of regulations and constraints, we virtually put all the burden of balancing out supplies and demands of credit on the price of credit, the interest rate. And that price must be high enough to discourage a lot of borrowing at times like these. We may not like it, but the fact of the matter is that many of the housing credit reforms of the past 10 years have helped accentuate the interest rate cycle.

There are some consolations in this. Our economy seems to be prone to such episodic attacks by the business cycle. They seem to be like the common cold or cancer: We have not yet developed adequate preventive medicine. All we have managed to do is to fight the disease once it emerges with some kind of bad-tasting medicine like curtailed federal credit programs, curtailed federal spending, and tight monetary policies.

Such anti-inflationary policies hit the building industry hard. The best that can be said for them as far as the building industry is concerned is that, if they do manage to hit inflation hard enough, they also may hold down the inflationary racheting up of prices year after year that is, in the long run, the insidious enemy of financial saving. And a good flow of financial saving, you remember, is a major nutrient of the construction industry. This may be one of those cases where short-run problems are part of a long-range alleviation of problems. But which is worse? I leave it to the Panel on Financial Planning for Improving Productivity to determine whether the cure is worse than the disease. We do know we are plagued with a good bit of both right now.

So much for giant economy size financial problems. The participants in the Forum planning workshops also put their collective finger on a number of company size financial problems that hold down productivity. Time and again it was emphasized that a major constraint to construction productivity is the inadequate equity cushion of many firms that leads them to extend themselves financially in ways that make them vulnerable to influences that cut their productivity.

Then we have the oscillating character of credit needs for the business. One set of swings in credit needs reflects seasonal changes. Another bigger set of credit swings reflects the ups and downs of our business cycle--swings in credit needs by builders, by buyers, and by suppliers. If we cannot either change the weather or eliminate the business cycle, individual builders are going to have to do much better than they have in the past in handling the financial consequences of these external forces in ways that do not play havoc with their own productivity. I would not be surprised if this is not an essential conclusion of our panelists. It certainly is a major challenge for them.

One of the difficulties in preparing for fluctuating credit needs is the striking imbalance between the financial capacity, the financial needs, and the financial incentive of the participants at various stages of the building process. Look, for example, at the contrast between builders and their suppliers. Many building materials and equipment manufacturers are much stronger financially than many of their customers, the builders. Are there ways to use that supplier strength to help support the builders and make everybody's productivity better? What is the incentive to the supplier to do just that? What will protect him if he does?

There is another kind of imbalance between the builder and the final buyer. Significant economies in future operating costs can be built into a facility at the cost of some added expenditures during construction. The reverse is also true. Initial capital costs can be cut in ways that will increase long-term operating costs significantly. We do not yet seem to have enough imagination to figure out how to divide these costs and benefits between builder and buyer in generally acceptable ways that give both enough incentive to achieve optimal productivity.

We also have a mismatch between the builder and buyer and the lender. In an inflation-plagued world, much of the financial gain from owning a building lies in either the appreciation of the price of the building on resale or in the escalating value of rent as inflation increases over time. On the other hand, much of the cost burden of owning a building results because lenders mark up the interest rate to try to protect themselves against the depreciation of the dollar. The former is an equity-type gain and the latter, a debt-type gain. Are there some sound operational ways to divide these gains and costs so that they better match the capacity of each of the parties to a transaction? For example, is there a mutually acceptable way to give the Bank of America an equity kicker in exchange for charging the builder or buyer a lower interest rate?

We also have heard again and again that financial planning in the building process is far less adequate than it ought to be for productivity purposes. Perhaps the Panel now will tell us how to plan a little bit better.

PRESENTATIONS

G. RALPH GUTHRIE

President
Urban Investment and Development Company, Chicago, Illinois

Urban Investment and Development Company is a wholly owned subsidiary of Aetna Life and Casualty, an example of the joining together of a major real estate operation with a major financial institution. All of Urban's activities are in the United States. In a broad sense, there are three on-going activities in the company. Urban owns a residential subsidiary that has an annual volume ranging from $35 million to $50 million. This company constructs single-family and multifamily residential properties in the Chicago area. Urban also has a wholly owned general contracting subsidiary that has had an annual volume during recent years of from $75 million to $150 million. Its activities involve commercial construction, primarily retail and office construction and often high-rise construction in the midwestern and eastern parts of the United States. Finally, and most significantly from a dollar and investment standpoint and because of the challenge presented, Urban is a real estate investor and developer and generally has from $300 million to $500 million of projects in progress in at any given time.

Urban is unique in this industry because of its diversity. Of Urban's multi-use buildings in existence or under construction, the most notable is perhaps Water Tower Place in Chicago, which houses more than 3 million square feet of retail, parking, residential, office, and hotel space in one building envelope. Urban currently is building a multi-use structure in Denver that will house a 620-room hotel as well as some 500,000 square feet of rentable office space.

Urban also is in the business of owning and developing office buildings, and it has a series of office buildings in Chicago, Denver, and Houston that range in size from 30,000 square feet to 3 million square feet, which happens in the latter instance to be a building under construction in Houston.

The largest single element of Urban's developmental activity, however, has been and continues to be regional shopping centers. This began in 1957 with Old Orchard Shopping Center just outside of Chicago. Today the company has six regional shopping centers in the Chicago area and a seventh under construction. This brings the firm's total shopping center holdings to something approaching 7 million square feet.

Urban also is in the hotel business. It operates the Ritz-Carlton Hotel, a part of Water Tower Place in Chicago. I already mentioned the hotel under construction in Denver. Two hotels are being planned for Boston and one for Seattle, which will be a renovation of the Olympic Hotel. Urban is in the industrial building business too. It began moving with some force in this direction a couple of years ago and now has completed or under construction some 2 million square feet of industrial space in Chicago, Denver, Atlanta, and the Baltimore-Washington corridor. These activities of Urban probably influence some of the remarks I will make today.

Construction is a boom or bust business. Thus, we must determine how to use that characteristic to improve productivity. The horns of a dilemma? Maybe it is. One element of the dilemma certainly is the money problem--the raw material of money. No matter what our product, we cannot create it without having the raw material of money. I am not an expert in money matters, but it is so important to our business that I must say a couple of things about it.

First, I am impressed with some of the recent measures that have enhanced the money supply to our various industries. Some new debt instruments have come into being--mortgage-backed securities, savings and loan money certificates, and the like. Some existing sources of funds have become newly available to us whether we are in the construction industry or the development industry. Some of us, including Urban, now are able to borrow through the commercial paper markets. Some members of our industry now are able to tap the Euro-dollar market. We also are beginning to reach the pension funds; they have moved rather slowly toward real estate and construction products, but they are moving and are an immense source of funds. Even the elusive foreign capital markets have begun increasingly to see the merit of what we are building in the United States from an investment point of view. I think there is to us a challenge to tap these new sources.

Now let me comment on interest rates, what money costs, and business cycles. In my mind, these are intimately entwined. To be blunt, I think that present interest rates are absolutely horrendous, absolutely wrong. They cut right to the core of all of our businesses. If we are businessmen, one thing we try to avoid is surprise. Experience teaches us we cannot avoid surprises completely so we eventually become accustomed to some surprises. Dealing with surprises within a given amplitude and degree is part of the challenge of management, but the increase in interest rates during the past year far exceeds those bounds. It was a major surprise, and I do not believe we should have to live with major surprises and hope we do not have to continue to do so.

Let us look at ourselves however. We also have some opportunities to do better. It all begins with quality of management. We really do not have qualified management in certain parts of our industry. I think we have been caught in the longevity syndrome in many places. In manufacturing industries, for example, people are elevated on the basis of longevity, but craftsmen and field superintendents are not automatically managers. I do not think longevity alone can be allowed to be the determinant of what constitutes the senior management of our companies. There is a tendency to avoid bringing in management specialists, but I think we should.

I do not believe that our business's planning and control methods are at all what they should be. General contracting and housing businesses in particular lack plans that deal with the total perspective of the business for the current and coming couple of years. Without such planning, how can you know what resources, money resources and people resources, you need to effectively and profitably run your business? In this respect, planning can be much improved.

I also find an almost complete absence of market analysis. Each of our businesses has some things that are clearly and truly marketable. What we ought to be doing is finding out what they are and who needs them and getting the two of them together.

One of our best opportunities to defeat the boom or bust problem is to better define our products and where we build them. This country is changing; the needs for our products are changing. During the past year in Chicago it has been almost impossible to sell new single-family homes. On the other hand, multiplex products have sold well in the face of a declining housing market. Our companies must have the ability to react to change and to new opportunities.

Geographically, two major opportunities are increasingly important. Sunbelt areas are growing at a substantial rate and I expect that to continue. America's cities are receiving greater attention, and I am pleased to say that Urban is a major contributor to and involved party in this increased attention. These are important changes. If we are going to run our business responsively, we must determine what the changes are and be a part of them.

We often hear about quality control in manufacturing and rarely hear about it in construction, but every one of us has experienced this scenario: The job is over, the money is collected, and the profit is booked. Then the explosion occurs--you've got to go back for a major fix. Why do we not have surveillance of the quality of what is produced? The cost of fixing is a multiple of the cost of doing it right the first time. That is a waste of money because it is going to come out of your pocket either directly or through insurance premiums.

Another waste area is cash management. To me, this is a simplistic thing. When I do my job, I want to get paid. I want it promptly. My money is expensive and valuable. When I warrant being paid, I want that money very rapidly. I want to make sure my people are exerting controls that make certain the money comes when it is due. Money is the raw material; I want to use it most effectively.

So we are a boom and a bust industry. But I think we have opportunities. One of them is to speak out on the debilitating impacts of these severe business cycles. Another is to improve ourselves.

OLIVER H. JONES
Consulting Economist
Oliver Jones and Associates, Manns Choice, Pennsylvania

As a replacement for one of the scheduled panelists, I can be considered to be the designated hitter for this group. The designated hitter is the guy the manager puts in hoping that he will not strike out. I plan to strike out. I plan to strike out at some of the shibboleths, beliefs, and complaints that you keep hearing in this industry. I have heard them for a long time, as far back as 20 years ago when I was with the Federal Reserve. I plan to say some things that are going to make you angry. I hope they will be provocative, and I hope you will think about them before you get angry. And, I hope you will think about them when you get home. Then you will be too far away to throw anything at me.

In a way, I will talk about Dr. Holland's giant economy size problems and how we get from there to the company size problems. First, however, let us get a couple of things out of the way so you will understand my definitions.

We are talking about productivity of capital. It cannot be very productive if you do not have any. So first you have to get it. Then you have to increase its size, husband it, and multiply it. What is capital? As Dr. Holland stated, it all comes from savings. There is no other place for it to come from except the credit manufactured by the Federal Reserve.

We must distinguish between two kinds of capital: Equity capital is the money that you put in as an owner. Debt capital is, of course, money that is borrowed. In the building business your equity position is a very important part of your ability to manage capital productively, to obtain debt, to negotiate with bankers on rates.

For example, this morning the interest rate is prime plus 3 percent and floating. Nevertheless, I know 15 to 20 percent of the people in this business are not paying prime. If they were paying a floating rate last year and, assuming they had the equity, they were smart enough to go to their banks and agree to buy the floating rate, because they wanted the money, but only if there was a ceiling. Today, those developers are living with a ceiling of 11 or 12 percent, which sounded outlandish last summer, but they are ahead competitively because they are not eroding their equity capital.

How closely are debt and equity related? How does equity determine your ability to obtain debt, to leverage equity productively? How do you build equity capital?

If you review the Horatio Alger stories, many of which are very honest stories, you will find that in the 1920s and 1930s and after the depression people started truly giant empires selling shoe strings on the street, saving their own equity, putting that to work, and making it grow. It is true that these people did not pay any taxes and that is part of their success. Although we do pay taxes today, the construction business has more tax incentives than any other--depreciation and tax shelters, the benefits given to the thrift institutions, the benefits to the borrower on his tax statement, both the interest rate and his real estate taxes. Relatively speaking, you are in a good position to accumulate capital in this business, much more so than almost any other business. I have none of those benefits in my business. Any capital I produce must come out of savings from current income. Thus, high taxes are not a satisfactory excuse for failing to accumulate equity.

Let us now get a couple of other things out of the way. One, business cycles--boom or bust, little ones or big ones--are always going to be with us. It is foolish for us to talk about developing a stable flow of credit into the home building or any part of the real estate and construction industry. It just is not going to happen.

For example, consider what happened last summer. It was decided, I assume by the present Secretary of the Treasury and the past chairman of the Federal Home Loan Bank Board, that, while pushing up interest rates to slow down the economy, steps would be taken to keep money flowing into housing. Thrift institutions were allowed to pay higher rates for six-month certificates with a $10,000 minimum. Thus, mortgage funds remained available but their cost multiplied.

Rather than reduce inflation, this gimmick increased inflationary pressures. At this point, borrowers are so sure of inflation, are so sure that real estate is the place to bet on inflation, that they are willing to pay higher rates. As a result, the thrift institutions, with low-rate, long-term mortgages still on their books, are accepting short-term deposits at increasingly higher rates. The squeeze that should have happened last June is happening now.

The thrift industry cannot escape disintermediation when credit is being restrained, nor can the building industry escape the boom or bust cycle. The industry uses a big chunk of the credit pie. Every year the residential mortgage market takes more credit out of the capital market than any other user. Accordingly, adjustment to credit availability has to take place there.

There are winners and losers in this process. During past periods of credit restraint and economic busts, we know that developers, real estate lenders, and mortgage bankers went broke. We also know, but tend to ignore the fact, that some people made money during the bust as well as the boom.

We typically complain about the bust when we talk about assembling the capital so that we can be productive. What did we do during the boom? How many developer-owners went out and put good projects together and made money, from which they could take out some capital for the coming bust? How many pyramided those successes and spread their equity over too many projects while increasing the number of Cadillacs they were driving? I cannot tell you how often I have examined financial statements of good developers with good track records that practically disappeared when the first stone was removed from the pyramid. The opportunity to develop capital in 1972 and 1973 was wasted in 1974.

We have looked at the enemy, and it is us. The outstanding example of this point is the Real Estate Investment Trust debacle of 1973 and 1974. There was no shortage of credit. Lenders were pushing credit on the developers. "Go out and build me something," they said. They wanted to put their money to work without worrying about whether there was a permanent loan behind it. Unfortunately, we did not have enough experienced developers, with or without capital, to handle the volume of credit that was made available for construction. We did not have enough lenders with the experience to make construction loans in the volume and types that were made. We had one great opportunity to develop capital. Instead, we simply inflated the leverage on equity and lost equity capital late in 1974.

We also use inflation as an excuse for the failure to develop capital. Inflation is going to be with us for a long time. I do not see anything coming along that is going to get rid of it. I am very supportive of what Paul Volcker did in October. Nothing done before then made any sense. We were in an inflated economy and everybody had begun to believe there would be no way out. Real estate looked like a good place to hedge against inflation.

Simply increasing the cost of credit was neither slowing down activity nor slowing down inflation. Both interest rates and inflation rates were increasing, but money was still available. The change came when the Federal Reserve decided to control member bank reserves. That should have been done a long time ago.

If the Federal Reserve succeeds in restraining credit expansion, a recession is most likely and, once again, equity capital in the construction industry will be threatened. I do not think we can look to the government for assistance this time.

Monetary and economic policy in this country is going to be made in Europe for the next six to nine months. Consider some facts. We now have more dollars abroad, claims on us, than the total money supply in this country. The OPEC nations really have not increased the cost of oil in 1979, not in terms of gold and not in real terms. They are simply responding to being paid in cheap dollars. The Germans have told us they will not give us billions more to support the dollar so we can push our inflation on them.

It is for these reasons that I believe U.S. policy is going to be dictated by Europe--not in a direct, literal sense, but we certainly will have to pay more attention to Europe than we ever have before. As a result, we can expect continued high levels of interest rates. If the recession bites, as it will, we will see a decline in interest rates but it will be modest. The inflation rate also will decline only moderately.

Let us get back to the boom and bust cycle. What are we going to do in 1980? If we demand more money from government, where is it going to come from? Are you willing to live with 18 percent inflation and a 20 percent prime rate in 1981? That is what will happen if you ask for more from Uncle Sam since the only way he can provide it is to print money.

In short, if you are going to get capital in this industry that you can manage productively, you are going to have to develop equity on your own side. You are going to have to develop as much as you can during the recovery. You are going to have to be astute enough to see the surprises coming and to slow down your operation to meet them. You are going to have to husband your capital during the recession periods.

ROBERT P. MARSHALL, JR.
Vice Chairman
Turner Construction Company, New York, New York

I am going to talk about the company size problem that Dr. Holland mentioned earlier. I am very happy not to have to talk about the broader picture since I am a specialist, not a generalist. I have worked all my life for a construction company, one that has been very fortunate. It has been through many boom and bust cycles, and I would venture to say that it probably will go through a few more in my lifetime and maybe even some during my period with the company.

Turner Construction Company is involved basically in the nonresidential building field. It is involved in a minor way in multifamily high-rise housing, but that is only a small part of its business. The segment of the market that Turner is interested in (i.e., nonresidential) was worth about $45 billion last year according to the Dodge Report. I think Turner is very fortunate that its put-in-place volume ran a little below $1 billion last year; it will be a little above $1 billion this year. One billion dollars flows through our cash register every year. That is a lot of money. A lot of it goes out too--about $975 million--to material vendors, to subconstractors, and to the company's production work force (laborers, carpenters, etc.). Of the $25 million left, about $21 million is used to cover general expenses, taxes, and other miscellaneous costs. Thus, the company ends up with a net income of about $4 million or about 0.4 percent of the original $1 billion. Half of the $4 million goes to stockholders while the other half goes into building up net worth. I mention these figures to illustrate that net profitability is very low. Although the picture is a little better if you look at it on the basis of stockholder equity; however, many banks look at our income statement and wonder why we are willing to work on that basis.

Contracting is also a very risky business. Turner has about 200 active jobs worth from more than $100 million down to $3 to $5 million. Of the 200 jobs, 5 or 6 are going to be bad apples. Those few apples, as we well know, can pretty well spoil the whole barrel of annual earnings. Thus, as one of the largest firms in the country in this business, we are and must be terribly careful and terribly alert. If we were not, we would not have been in the same business, at the same shop, under essentially the same management for 77 years. There are not too many other large construction firms that can say that.

How do we cope with low profitability and high risk? The answer is the subject of this segment of the Forum program--financial planning. The one outstanding characteristic of the company since the 1940s has been a real sense of financial planning. Before that time, we were bricklayers and carpenters, the fellows out in the mud and dirt. Mr. Guthrie is right when he says we have to change the kind of people that are running our kind of business. And we have changed. We had to learn to run a tight ship, and I think that is what a lot of other people in the contracting side of the business still have to do. This has to do very much with productivity, the subject of this Forum. There are many tens of thousands of contractors in this country. At least 8,000 general contractors are members of the Associated General Contractors of America, and there are many trade contractors and small wheelbarrow-pushing outfits in the contracting business. Many of these contractors are going to disappear or be a very heavy drain on capital because they do not pay attention to financial planning.

At Turner, financial planning starts with tight budgeting on a yearly basis followed up by adequate monitoring of our 20 profit centers. This approach is certainly nothing new to any industrial firm or financial institution, but it is truly unusual in our business to run that kind of a tight ship, where you have a definite budget and definite responsibilities to stick to budgets. Many subcontractors do not do this, and we find them going under. Some of the largest in the country have gone under within the past year because of poor financial management.

We have 20, 30, or 40 joint venture partners who are leaders in their part of the country or in their particular specialty. We are constantly amazed at the quality of financial planning, financial control, and awareness of what is going on. Most construction companies are indeed seat-of-the-pants operations. This becomes more obvious to us every day as we move into new areas of the country and see how some jobs are working.

I hope that you can see a definite connection between my remarks and productivity. For as long as I can remember, some 40 years in the business, we have been using labor as the fall guy for all of construction's problems. Cost, time, and quality disasters most often are laid at the feet of the people who are actually out on the job site working. Many of them work under union conditions and many of them work under open shop conditions, but in many cases the buck goes back to them. I think it is very convenient for management to lay all its ills at the feet of someone else. This practice probably was much more acceptable to the rest of the workers when the boss was a lot "closer to the wheelbarrow" than he is today.

I believe we must improve the level of management in the construction business, and I believe that this should be one of the major points that are mentioned in the proceedings of this Forum. We hear a lot about training people at the working level, about Equal Employment Opportunity, and about the great mass of people that we could bring in to the industry, but even if we can bring in all those people, we are going to be in trouble if we do not raise the level of management.

Dr. Siegel commented earlier that systems alone can accomplish nothing. I heartily agree. We have many systems; what we have to learn is how to use them effectively. Turner is as subject to criticism as others for spending considerable money developing systems but not adequately utilizing those systems. However, to benefit the entire industry, many more have to learn how to use the marvelous tools that we have at our disposal.

We have to learn how to care for the billions of dollars that go through our tills every year. We are going to have to be able to project better cash flow requirements so that we can advise the owner more accurately concerning timing. Once we have the money, we must handle it more effectively and pass it along quickly to the material vendors and subcontractors so that they do not have to plan on financing their material costs before getting their cash back. We make every effort to have a reputation for paying according to contract so that the subcontractors know what to expect when they give us prices. We must use effectively any funds lying dormant.

We also must learn to schedule our year-to-year work load to alleviate the boom-bust or hire-fire syndrome. Turner has done that to a substantial degree; we are not perfect but we certainly are a lot further along than many others. We look forward to using another thing that was mentioned earlier--marketing analysis. We used that technique very successfully four or five years ago and were able to prepare for the overnight disappearance of the commercial market. Because we knew what was going to happen, we started to energize a marketing program in the hospital field and, at one point, were doing 50 hospitals. We were not doing this entirely by choice; there simply was no commercial work available. We also must tighten the links of communication between developer, financial source, user, designer, and constructor so that our respective activities are more closely meshed and therefore more productive.

I strongly urge that this Forum emphasize clearly the need for more rounded development of tomorrow's construction leaders. It is imperative that they be better fitted to work as managers in the fullest sense of the word.

THOMAS J. FLYNN, JR.
Senior Vice President, Real Estate Loan Department
Bank of America, San Francisco, California

I would like to cover three basic subjects: the California situation (because in social and political issues it is something of a precursor of activity on the national scene), the availability of institutional capital for both construction and long-term real estate loans in an inflationary economy, and recommended courses of action based on the above.

With regard to the California scene, my remarks will focus primarily on the housing situation since it is somewhat unique because of the inflated value creation we are facing out there. Some things that happened just yesterday will be of interest. We had a statewide election in California yesterday, and two statewide initiatives were considered. Proposition 2, a usury initiative, went before the people of California for the fourth time; it passed, eliminating the archaic 10 percent limitation on nonexempt lenders that California has had since 1934. Proposition 4, the follow-up to Proposition 13, also passed. It was initiated by Mr. Gann who was Howard Jarvis' silent partner. What it does is to put some rather stringent limitations on the spending abilities of local government. Proposition O, an anti-high-rise initiative sponsored by the citizens of San Francisco, and Proposition R, which further tightened up the rent control established by initiative earlier in 1979, were defeated at the polls.

On the economic scene, our prime rate is 15-1/4 percent; the Federal Reserve rate is gyrating wildly around 15 percent; and large certificates of deposit are costing national banks, after the increased reserve requirements, about 16 percent. Money market certificates are now costing savings and loans 12.193 percent whereas three weeks ago that figure was 10.66 percent; as a result, most savings and loans in California have shut their doors to new business. The best rate for a conventional single-family mortgage (80 percent, fixed rate, for loans over $50,000) is currently 13-1/2 percent and two points; it was 11-3/4 percent three weeks ago. Construction borrowers are paying about 18 percent. We would like to cap the interest on these loans, but first we would like to have a ceiling on what we pay for money. We have very few capped construction loans. The Federal Housing Administration (FHA) raised its maximum home loan rate to 11-1/2 percent two weeks ago. That sounds okay, but the present discount to the Fannie Mae market is between 10 and 15 points. We are currently at 10.

This is a timely conference. As the Forum brochure states that "construction activity has been consistently plagued by the instability of economic cycles." I must say that this has to be one of the most anxiously awaited recessions that I have ever had the opportunity to experience. Maybe it is here. If it is half over, I hope the second half is not any worse.

Let me now say a little more about post-Proposition 13 California. I voted for Proposition 13, and I have seen little happen since its approval that worries me. However, the full impact is still to be felt in terms of the ability of local communities to finance infrastructures, housing, and other local services.

On the housing side in California, demand remains exceptionally strong. We were concerned when home loan rates reached 10 percent, but the demand for mortgages is still strong. Last week we had 1010 applications on houses for appraisal in the state and that is sustainable; however, as I already said, the savings and loans are out of the market. We expect 200,000 housing starts in 1980, and we should have that many in 1979. This compares with 233,000 starts in 1978.

Thus, higher prices have not yet stifled housing demand in California. In-migration is up, and our demographics are favorable with many in the 24 to 35 age group first entering the home buying market. Real demand for housing in California, both multiple- and single-family, is about 300,000 units. The problem that we face is one that you all face: The cost of new and resale housing is obviously outstripping real personal income. In California, the median house price in August 1979 was $85,369, a 19.4 percent increase over August 1978.

What has Proposition 13 done? Thus far, it has increased the cost of housing and of manufactured construction projects because local communities have turned to increased processing, sewer hookup, plan check, and other fees in lieu of park and school fees to offset lost revenues. New housing development is discouraged because of increases in infrastructure costs. It is roughly estimated that these fees can add from $5,000 to $10,000 to the cost of a newly manufactured dwelling in California.

I will not comment on environmental pressures. There are two sides to that issue. Nevertheless, such pressures have not reduced land cost.

The prospect of rent control has severely depressed multi-family residential starts in California. Apartment vacancy rates in some areas are below 1-1/2 percent.

What about the availability of institutional capital for construction and long-term loans? At a recent luncheon, Dr. Mark Reidy noted that the total residential mortgage debt in this country held by individuals, institutions, and the secondary market is in excess of $900 billion. It exceeds the national debt! As of October 6, disintermediation became a reality. Fixed-rate 30-year mortgages have become singularly unattractive institutional investments, and usury limitations have been very harmful in terms of discouraging movement of capital across state boundaries.

Now for some courses of action. As far as capital formation and mobility are concerned, I think that we need to encourage savings with a federal income tax credit exempting at least the first $100 of institutional savings interest for an individual and the first $200 for a joint return. The savings rate in this country is below 5 percent; it is abysmal. There are few incentives to hold dollars and I believe such a credit could help. Although I do not know exactly how much such an approach would cost, I think it would be less than some other federal subsidies.

I also believe we need a national usury pre-emption bill that will cut across state usury initiatives. Perhaps this would come under the jurisdiction of the Interstate Commerce Department. Chicago, for example, has been starved for money and, until yesterday, so was California. In Tennessee, people were leaving Memphis to go across the border to Arkansas because of archaic usury laws. Such laws have impeded the flow of mortgage and construction capital even though it is a national capital market. Why should California pay more to exempt lenders like ourselves? If California law permitted free access by institutional mortgage investors, office building developers would be paying in the 11 percent range for their permanent money. Bank of America's best rate on a credit income property deal today, immediate funding, is 13-1/4 percent and two points. That is not where the market ought to be.

New mortgage instruments also are needed. We have pioneered variable rates at the Bank of America. We have both a six-month variable rate program and a five-year rollover plan. I think the rollover plan is the wave of the future. Of the Bank's $2.4 billion new residential lending total in 1978, 26 percent of the borrowers opted for the variable rate plan.

The next type of alternative mortgage instrument needed is a graduated payment mortgage (GPM) instrument. The FHA has its 245 Program now, and the Bank of America is going to introduce a GPM plan early in 1980 that will help first-time home buyers to afford housing. A GPM program allows for lower initial year payments. For example, a $75,000 borrower with a conventional 10 percent downpayment would have to make principal and interest payments of $617 a month. Under the Bank of America's graduated payment parameters, his monthly payment would be $515 per month for the first year. This instrument is not altogether attractive to us. It offers lower cash flow, and we cannot sell it easily. Hopefully a secondary market will emerge.

The FHA also has announced a co-insurance program, and the Bank of America supports this program. It involves a shared risk, 90-10, between the lender and the FHA and allows for a higher ratio loan for inner-city lending.

Last, I think that the FHA and the Veterans Administration (VA) finally should get away from mandated interest rate maximums on insured mortgages. A 10-point discount being passed on in extra cost to the home buyer is extremely inflationary. This is, of course, a political issue, but I think that it is in the best interest of the consumer and the home buyer to let the FHA and VA rates float with the market.

WILLIAM J. CALDWELL
President
Caldwell Equity Corporation, Troy, Michigan

My firm is involved extensively in both commercial and residential construction; however, my remarks will pertain to residential construction. Our company builds both attached and detached housing for sale. Although primarily a Michigan-oriented firm, we also build in New York, New Jersey, and Florida.

The housing industry is the most volatile and cyclical facet of the building industry. Compounding this problem is the fact that residential builders generally are small entrepreneurial type organizations. There are a few very large national building companies (e.g., U.S. Homes, Ryan Homes, Centex, and Lincoln Properties), but their annual volume represents only a small percentage of the 1.7 to 1.8 million starts that we will have in 1979.

Most of today's builders are survivors who cut their teeth during the 1968-69 crunch and the bad recession and over-building problems during 1973-75. In 1975, the industry was distressed by very large product inventories, excessive land commitments, and high interest rates on large leverage construction loans. Housing starts declined 60 percent in 1974-75. We could only wonder how something so "REIT" could go so wrong. A modification of Murphy's Law seemed to apply: Whatever hit the fan was not evenly distributed.

Most housing companies now are finely tuned, well disciplined, and basically conservative organizations. Financial and planning controls are essential. Cash flow is more important than balance sheets. I think the Turner and Urban organizations understand this as well as anybody. To think that an organization (Turner) can put through billings of $1 billion in one year and keep only 0.4 percent profit. The housing industry has structured itself differently now. We have protected ourselves and built in safeguards, both legal and financial. We deal differently with lenders. We talk about limited partnerships and limited liability, and we have learned to gear up or down quickly.

The money crunch seems to occur every three to five years; since we can count on the consistency of this from all our past history, it would just seem logical that we program this into our projections and plan for it. Thus, as we enter this 1979-80 recession period, the housing industry seems to be in pretty good shape compared to 1973-75. Our lending institutions reacted early and proceeded with extreme caution. We do not expect large product inventories, and we should be able to withstand a substantial slowdown.

Earlier speakers indicated that productivity in our industry is influenced by both external problems, those areas of concern at a national level, and by internal problems, those more closely controlled by our own corporate structures. On a day-to-day basis, we really have very little control of the external problems. We can help ourselves externally by dealing with the political entities involved in our industry through the elective process and we can deal with our various trade organizations at national levels, but because of its scale and size, a housing company can use only its own best estimate for planning projections. We can evaluate the predictions of various economists, of financial institutions (e.g., Dr. Evans of Chase Econometrics), and of our national associations, but ultimately it must be our judgments and decisions that determine which assumptions we use in our long-range planning projections.

Today, not even the experts or the federal government can agree on the answers or predictions of the future. In October 1978, gas was selling in Detroit for $0.70 per gallon, mortgages were at 10 percent, the prime rate was between 9-1/2 and 10 percent, and the annual inflation rate was approaching 8.5. Today, businessmen and economists alike find themselves confused. The annual inflation rate is 13.2, gas costs $1.04 per gallon in Detroit, mortgages are over 14 percent for those lucky to find them, businessmen are paying more than 15 percent to borrow at the prime rate, and construction loans are running between 17 and 18 percent.

What should you have done internally to effectively plan your corporate strategy for the balance of 1979 and 1980? This is an area in which the housing business is in a good position to control its destiny. The suggestions I will make are in no particular order of priority, but many in our industry initiated these strategies early in 1979. Most of us bought end loan mortgage packages. Unfortunately, most of them are expiring and will be used up by the end of this month. We also can begin enhancing our cash position and building liquidity. I would not build a speculative inventory of housing today.

I would emphasize development of a capital base by seeking joint ventures or by making other equity arrangements. My company has been involved in joint ventures since 1973. We currently have such arrangements with a financial institution which you heard about earlier, with a general contractor on a commercial project, with a Detroit-based Fortune 500 company on a large housing project, and with customers of a European bank on a project in Florida . Such arrangements are available to any company in our industry that has credibility and a good track record. I would recommend that you use any of those avenues to initiate a larger capital base during this period.

To summarize, let me note again that the residential housing industry is very sensitive to the conditions of the national economy. It is made up primarily of small businesses with entrepreneur-type managers. It is neither a precise nor an exact business; however, tight financial and construction controls are a must in this very demanding period of our economy. It is certainly a time to run a tight, well disciplined ship and to measure the risk-reward of each project in a very conservative manner.

History has shown that housing and development companies with good cash positions have made substantial profits during recession periods. We can look forward to the same opportunities during the months ahead. It will be a period during which the strong will get stronger, and many of the weak will go out of business. The next 12 months will indeed be most challenging.

DISCUSSION

QUESTION: What can the building and construction industry do with financial planning to improve productivity as opposed to profits? Profits do not result only from increased productivity. What kind of financial tools can best be applied to improve productivity?

MR. MARSHALL: The kind of quality management I talk about benefits the owner and the user. Competition is necessary to make this work. Fortunately for the United States and unfortunately for us as a company, we have lots of competition. By using a total management approach to the construction business that involves, for example, a very active cash management program and the prompt payment of debts, we can increase our ability to be more competitive in making a deal. We do not like low profitability. Nevertheless, I assure you it is not untypical of the firms in our business. With regard to specific tools, I think the answer is a total management approach to the conservation and utilization of the amounts of money that an owner commits to a particular project.

MR. JONES: Let me give a specific example. Consider the large home builder or even an office building developer. If he has sufficient net worth, he can get credit. In recent years, he has been buying ahead, putting materials in a warehouse near the job, and moving them on the job when he needs them. He is using his capital more efficiently by moving materials to the job in a timely way than by having people standing around, laying them off, or postponing subcontractor activities. Now let us see if we can use that equity capital even more effectively. Suppose the developer entered into a deal with the material supplier. The supplier would sell the material to the developer today at today's prices but keep it in his warehouse and deliver it precisely when the developer wanted it. The developer would give the supplier a day or two for planning and would pay for the material supplier to keep the materials in his warehouse. The supplier's credit is much better than the developer's; therefore, the carrying cost is going to be much less. This is one possible way of getting a material supplier to work with you.

MR. MARSHALL: Just a small warning about that approach. We have done exactly what Mr. Jones described and have found that we must be very careful about the clarity of title and possible lien rights of one kind or another. Be equally careful about the details of insurance coverage.

QUESTION: Mr. Flynn, you mentioned real demand as being 300,000 units. How did you define real demand? What effect do you think graduated payment, invariable-rate mortgages will have on that demand?

MR. FLYNN: The 300,000-unit demand for residential housing in California was developed by the state based on demographics, existing housing stock, housing going out of use, and the other factors that bear upon it. Actually, the estimate ranges from 280,000 to 300,000 units. With respect to the graduated payment mortgage (i.e., lower but increasing payments during the first five years so that the borrower does not even make a sufficient payment to service the interest and then a level payment for the last 25 years), the Federal Home Loan Bank Board estimated two or three years ago that such mortgages could put an additional 2.5 million households presently priced out of the market into homes. The Bank of America's marketing research department has estimated that we could transact 10,000 loans annually if we capture a good share of the market through the vehicle of the graduated payment mortgage. I do not consider the variable rate to be a real stimulus to home buying. Twenty-six percent of our conventional residential mortgage borrowers last year opted for the variable rate. At that time we had a 1/4 percent lower rate available to them as an incentive. I am fearful, as is the Congress, that lenders some day may just stop making fixed-rate, 30-year home loans. As an institutional investor, I think such loans are a singularly bad investment in these times. We are considering very seriously going to a variable-rate mortgage on a mandatory basis for income property.

QUESTION: Will interest rates will decline in 1980?

MR. JONES: They are going to decline but not significantly.

MR. FLYNN: I brought along a 60 percent probability forecast that might be helpful. It assumes a reasonable recession. The quarterly averages, prime rate, for 1980 are as follows: first quarter, 14-1/4 percent; second quarter, 12-1/4 percent; third quarter, 11-1/4 percent; and fourth quarter, 10.67 percent.

DR. HOLLAND: That forecast may be some of the best news we have heard.

MR. CALDWELL: The National Association of Home Builders' economists originally forecasted that there would be 1.5 to 1.6 million starts. Yesterday, they announced that starts would range from 1.065 to 1.400 million. They now have developed a formula based on what happens to the prime rate. If those of us who predict only a modest decline in interest rates are right, their formula would produce the low. This year there will be 1.7 million starts so that would be a real recession. If Mr. Flynn's forecast is correct, there probably will be 1.3 to 1.4 million starts.

MR. GUTHRIE: If the interest rates decrease as Mr. Flynn's forecast indicates, their impact on housing is going to be significant. Even with a favorable movement in the rates, I think it is extremely optimistic to expect more than 1.3 million starts.

MR. JONES: Some of the panelists were talking about which economist's projection to use and the need to use your own judgment in making decisions. There is no question about it: the businessman must use his own judgment. After all, it is up to him to determine what to do. When looking at forecasts and considering which one to use, pay attention to the assumptions that are being made and not to the numbers. Watch those assumptions as time passes. When they change, adjust the numbers and make your decisions. My assumptions were very clear: that Mr. Volcker's policy is going to stick because Europe is going to make it stick and that neither the Congress nor the President is going to make any significant change in policy despite the quadrennial election year. If that works out during the next few months, I think you will see a quicker change in the prime rate. On the other hand, unless the inflation rate decreases very substantially, the prime rate will not go down very far.

QUESTION: Since you believe that the Federal Reserve has taken the proper steps, what do consider to be the milestones that will indicate that the medicine is working?

MR. JONES: The Federal Reserve announced appropriate intentions on October 6, 1979. We will have to watch events to see if they are carried out. The first test will come in January 1980. If the Federal Reserve absorbs the return flow of funds to bank reserves that traditionally follows the year-end holiday buying season, it will be carrying out its stated intentions. If it does not, the excess reserves allowed in the banking system will tend to lower interest rates and encourage foreign holders of dollars to believe that inflation will worsen in the United States. This, in turn, will cause another run against the dollar and will force the Federal Reserve to take more stringent monetary actions. What does this have to do with the productivity of capital? As I noted in my formal remarks, the developer can use his equity capital productively, sometimes too productively through leverage--borrowing many times his capital base. What needs to be accomplished, therefore, is to accumulate and to conserve equity capital. In the situation described in this question, the emphasis must be placed on conservation of capital by avoiding overexposure and excessive leverage and by lining up credit resources in advance of serious restraint.

QUESTION: What would you have recommended in 1973-1974 to build capital in the construction industry?

MR. JONES: I would have recommended a position opposite to that taken by most developers. The early 1970s was a period when equity capital could be leveraged substantially and, to a point, safely with debt. It was the halcyon period of the Real Estate Investment Trust. It was a period when equity capital could be and was accumulated. As the period progressed, developers overextended their leverage and their ability to perform. Few recognized their own inability to perform on all of their projects, let alone the general overbuilding of the period. As 1973 began to show signs of stress in troubled projects, few developers saw fit to conserve the equity capital accumulated in the early 1970s and continued to expand their operations. The few developers who responded to conditions were able to complete their projects and conserve rather than lose equity funds. They were on the right path. It is difficult to run against the grain and husband equity resources when every competitor is reaching beyond his equity and managerial resources, but it is far more productive.

QUESTION: How do you view Mr. Flynn's recommendations for encouraging saving through tax incentives?

MR. JONES: Without a specific proposal in hand, only a general reply is possible. If the tax incentives increase the federal deficit, they will contribute to inflation, decrease the value of savings, and encourage spending. If the tax incentives are provided within a noninflationary fiscal policy (i.e., reduced spending), savings will accumulate. As long as investors are willing and able to put the savings to work, such a change would be constructive. On the other hand, the specific proposal before the Congress (i.e., a $200 exclusion from taxable income for deposit interest earnings per couple) will do very little to attract savings to thrift institutions. This amounts to about a 1 percent increase in the rate of return to depositors in a marginal tax bracket as low as 25 percent, less after the deposit exceeds $4000. In today's high interest rate environment, this is a tax break for small savers who cannot or will not move their funds. To all others, the alternatives available--all above 10 percent--are still far more attractive than a thrift deposit.

Session III

GOVERNMENT AND PUBLIC ACTION FOR IMPROVING PRODUCTIVITY

PANEL OVERVIEW

MURRAY L. WEIDENBAUM (Moderator)
Director, Center for the Study of American Business
Washington University, St. Louis, Missouri

The public does not understand the tremendous extent to which government is the problem rather than the solution in the building business because of the increasing extent to which government regulations delay construction and increase costs. The public sees only that private builders are raising prices; many citizens do not connect their strong pressures for an array of environmental, safety, and energy standards with the resultant cost increases. A great deal of education is needed on this point.

One of the basic causes of the sharp increase in the price of new homes is the rising array of government regulations facing land developers, home builders, financial institutions, and others involved in the housing industry. To clear the air at the outset, I will state that this is not going to be an uncritical attack on all efforts of government to regulate the private sector; rather, it will be an evaluation of the impacts of regulation on housing. The evaluation will conclude with some suggestions for improving the status quo in business-government relations in this important segment of the economy.

Lest I be misunderstood, let me state the obvious: Government regulation often has yielded important benefits--for example, less pollution, fewer product hazards, reduced job discrimination, and other socially desirable objectives. It also should be realized that government programs generally are established in response to rising public expectations about business performance. Nevertheless, the worthiness of social objectives should not make specific methods used in attempting to achieve them totally immune from criticism. I am sad to report the almost instinctive negative and hostile reaction, especially on the part of many of the so-called public interest groups, to anyone who even questions any of the specific means that are used for social regulation. I find it unfortunate to have to remind these enthusiasts that only in a totalitarian society does the end justify the means.

At first blush, government imposition of socially desirable requirements on business through the regulatory process appears to be an inexpensive way of achieving national objectives. This practice apparently costs the government very little and represents no significant direct burden on the taxpayer. Government regulation sounds like the proverbial free lunch. But, as we have found out, the public does not escape paying the cost. Every time, for example, that the Environmental Protection Agency (EPA) imposes a more costly (and perhaps less polluting) method of construction on any firm, the cost of the firm's product to the consumer will rise. Similar effects flow from other regulatory efforts, including those involving product safety, job health, and equal employment (and credit) opportunity.

These higher prices represent the "hidden tax" of regulation that is shifted from the government to the consumer. It is not inevitable that every regulatory activity increase inflationary pressures. When regulation generates social benefits (e.g., a healthier and, thus, more productive work force) in excess of the social costs it imposes, inflationary pressures should be reduced. But if the costs are ignored and the focus of public policy is only on the benefits, it is almost inevitable that regulation will be pushed beyond the point at which the benefits equal the costs and into the zone of overregulation. Overregulation, to an economist, is not an emotional term. It is merely the shorthand for situations where the costs imposed by regulation exceed the benefits from the regulation.

The basic point of this presentation is that the regulatory process can be revised to derive, at lower costs, many of the same benefits that are now achieved. Before we turn to the subject of change, however, let us first examine more closely the major economic effects of regulation of home building.

Government regulation increases the cost of new homes in many ways. It drives up land and land development costs. It increases the number of expensive building code features to be incorporated. It raises the overhead expenses of real estate and financial institutions, and it increases financing costs due to project delays.

In recent years, there has been a rapid expansion of regulation affecting housing by all levels of government. The newer federal regulations cover a wide range. They include standards for water quality, pollution discharge, and dredge and fill operations; sanctions against localities that do not restrict developments in flood-prone areas; requirements for state and local governments to regulate activites that pollute the air; and regulations affecting closing and settlement procedures and the extension of mortgage credit.

Many states have extended their regulations affecting housing development. New types of rules include those governing building in "critical" areas such as wetlands, floodplains, and shorelands. States also have enacted measures to control erosion, stormwater runoff, and water and air pollution and to require environmental impact statements.

At the local level, the major change has been toward a fuller and more systematic use of traditional land-use control techniques. These are often supplemented with such new departures as development timing and rate controls, higher required contributions of land or facilities from developers to the local government, and special standards for marshes and floodplains. Moreover, so-called growth management has become fashionable in many areas. Such limitations on the supply of developable land, no matter what their motivation, force up the price of land and of homes generally.

From the viewpoint of the individual builder, government controls start when land is purchased and extend through the entire construction process and into the point of sale to the home buyer. If environmental or growth controls exist, the builder has to conduct a market survey to show that his development will be needed and will meet local standards. There also are geological and engineering reviews to determine what can be built on the land. The developer many have to retain the services of such specialists as land planners, civil engineers, soil and geology experts, zoologists, biologists, and traffic engineers.

Before construction begins, the builder must obtain a variety of permits and be subjected to numerous government reviews. Here is a sampler: approvals for the master land-use plan, environmental impact reports, subdivision plans, grading permits, building permits, and plumbing and electrical permits. In addition, the builder may be required to provide or pay fees for flood control, fire protection, schools, parks, open space, access roads, drainage, water, electricity, gas, sewers, landscaping, traffic signals, and signs.

Should the project be located in a coastal zone, the entire process becomes even more complicated and, thus, more costly. The typical delay when a coastal zone commission is involved has been estimated at three months, but it can take much longer. One project in Santa Monica, California, came under the jurisdiction of both the city and the California Coastal Commission. It took three years to obtain all of the required government approvals. While the project was thus delayed, construction costs increased by 60 percent and the sales price, by 30 percent.

Unfortunately, it is not difficult to find outrageous cases of regulatory delays. A single Boy Scout doing an ecology project was able to bring construction to a halt on a 200-unit condominium project in San Francisco, California. Bernard Frieden of the Massachusetts Institute of Technology points out that the regulatory review process in the housing area is highly political and that the people with the greatest stake in its outcome--housing buyers--play no part in it.

Let us try to measure the impact of the regulatory environment on new housing. There have been several efforts to qualify the growing costs that result from the rising array of regulation of home building and of housing activities. Several studies have examined the adverse impacts of overly stringent or outdated building codes. In a study at Rutgers University it was estimated that such costs are between 5 and 10 percent of total unit costs. Two Yale University economists have done some interesting work on identifying the extent to which building codes serve as barriers to innovation. They point out that the "bewildering variation" in local regulations can bar potentially profitable innovations in some areas. This reduces the size of the market for technical change in the home building industry, with a negative effect on the incentive of building material suppliers to perform research and development in this field.

More fragmentary, although intriguing, estimates are available of the indirect costs of regulation to the home buyer. Giving borrowers the booklet required by the Real Estate Settlement Procedures Act is estimated to cost $0.35. Completing the forms required by Truth-in-Lending legislation "costs no more than a few dollars." The National Association of Home Builders states that financing and carrying charges for home building amount to between $10 and $18 per day per lot. Thus, using the low end of the range, $10, a 6-month delay adds $1825 to the cost of each new home.

Speaking of delay, I found particularly fascinating a newspaper article quoting an undersecretary of the Department of Housing and Urban Development. The federal official recalled a conversation that he had with a county commissioner when he was a builder in Florida. The county commissioner was explaining his "pinball" technique for protecting the environment: "When a builder comes in with a certain project, I just bounce him around from one department to another."

Several comprehensive estimates of the cost of housing regulation have been prepared. A study conducted for the Colorado Association for Housing and Building found that changing regulatory requirements and practices had added between $1500 and $2000 to the cost of the typical new house built in that state between 1970 and 1975. The added cost consisted of higher water and sewer tap fees; increased permit fees; greater school and park land dedication requirements; and new mandates for wider and thicker streets, fences, underground storm sewers, and environmental impact studies.

A study in St. Louis County, Missouri, of the increase in lot development and home building costs during 1970-1975 indicated that the expense of meeting new government requirements came to between $1600 and $2500 for a typical 1600 square foot house on a 10,000 square foot lot. The new governmentally imposed requirements included street lighting, greater collector street widths, higher permit and inspection fees, added features to electrical systems, and smoke detectors.

A Rutgers University study of 21 residential development projects in the New Jersey coastal zone estimated the direct regulatory expenses for a single-family house at $1600 during 1972-1975. The costs covered some 38 separately required permits, including preliminary plat, performance improvement bond, sewer plan, tree removal permit, final plans review, drainage permit, and coastal area facilities permit.

One study that has received considerable attention was prepared by George Steinlieb, Director of the Center for Urban Policy Research at Rutgers University. He estimated the cost of meeting governmental regulations that exceeded what he considered to be "minimum health, safety, and welfare considerations in the provision of housing." As a result, his figures are higher than those of the other studies. He assumes, for example, that any required lot size above 1/4 acre created unnecessary costs for the home buyer. Overall, he estimated that excessive regulations on the average new home priced at $50,000 cost the buyer $9844 or nearly 20 percent of the total.

Professor Frieden also pointed out an indirect but rapidly growing cost of housing regulation. Some government jurisdictions are using various permits and other regulatory requirements to shift a portion of their existing costs to new housing development. This is a very grey area. There is some justification, of course, for assessing new developments--the added costs that they impose on a community. However, the computation of those costs is not easy, and there is the inevitable temptation for the jurisdiction to be extremely liberal in estimating the costs that it will impose on new home owners who are not currently voters in that locality. As Professor Fred Case of the University of California at Los Angeles has written, "There is no argument that controls have increased the initial and operational housing costs. There is considerable disagreement as to the levels of these costs and the extent to which they are 'legitimate'." That is a very fair evaluation of the situation.

What can be done to improve the status quo? Before we tackle that question, we need to remind ourselves that important and positive benefits have resulted from many of the government's regulatory activities. These government programs were established in response to a surge of rising public expectations about business performance. Thus, reforming government regulation involves striking a balance among many laudable objectives. It is not a search for villains. Indeed, the magnitude of the unresolved problems in the regulatory area requires efforts by many groups--government, business, academic researchers, and the various interest groups.

The basic task of government in the regulatory reform area is not to be preoccupied with either technical measurements of benefits and costs or administrative procedures, although good can be achieved by some sensible changes. More fundamentally, government leaders--at the federal, state, and local levels--need to take a dramatically different view of the regulatory mechanism than they now do. Rather than relying on regulation to control in detail every facet of private behavior, the regulatory device needs to be seen as a very powerful tool to be used reluctantly and with great care and discretion. A good deal of judgment is required in sorting out the hazards that are important to regulate from the kinds of lesser hazards that, in Charles L. Schultze's words, can best be dealt with by "the normal prudence of consumers, workers, and business firms." When the device of regulation is relied upon, the emphasis should be placed on identifying the least costly and most effective means of achieving social objectives. To state what is obvious to an economist but so often ignored in more popular discussions, you do not protect the consumer by punishing business.

The basic task of conducting business successfully in a regulated environment is extremely difficult. Business firms need to respond to rising public expectations for safe products produced in a healthy work environment, free of discrimination. To the extent that businesses respond voluntarily--and a great many do--the pressures for government intervention may subside. Yet, I do not advocate a passive role of automatically agreeing to every demand on the part of each interest group, public or private. Those demands that do not make sense should be opposed--lawfully and strongly--and more sensible alternatives should be developed and presented. Rather than vague speechifying on the evils of big government or the glories of the free enterprise system, business needs to concentrate its efforts on more effectively communicating the specific impacts of regulation on its production, sales, employment, and prices so that more balanced laws and regulations will be enacted.

Let me issue a word of warning based on experience. Do not play the role of Neanderthal. Simple-minded opposition to each and every government regulation is so patently self-serving that it is ineffective; in fact, it is counterproductive. The public believes that there are environmental, product safety, and discrimination problems. Some of them even realize that there is an energy problem. The serious question is how can we make that array of government regulation simultaneously less onerous and burdensome and more effective.

A little humility might go a long way in reducing the shrillness of many of the representatives of the so-called public interest groups. It is no simple task to identify the public interest in any specific issue of public policy. As a former participant in government policy-making, it is apparent to me that good policy-making is not a search for the villains. Rather, it consists of properly balancing and reconciling a variety of bona fide interests. This is far more difficult than merely choosing in a simple-minded fashion between "public" or "consumer" interests, which are presumably good and to be endorsed, and "special" interests, which are presumably evil and to be opposed.

Sensible public policy will not respond exclusively to industry's gripes about the costs of regulation or uncritically to the public interest advocates and the regulators who see only the benefits of their actions. One of the most serious problems in the home building area is the fact that the home buyer--who ultimately bears the cost of regulation--is not represented in the regulatory process. In fact, many of the proposals of the self-styled public interest groups would result in greater burdens on the home buyer.

Unfortunately, there seems to be a parallel between generals fighting the last war and academics researching issues of public policy. Whether I speak to business executives, labor union representatives, public interest groups or government officials, I find that their key concern with government regulation is in the newer cross-industry type of regulation typified by the Environmental Protection Agency, the Occupational Safety and Health Administration, the Equal Employment Opportunity Commission, the Employment Retirement Income Security Administration, the Consumer Product Safety Commission, and their counterparts at state, county and municipal levels. Yet, my academic brethren still seem preoccupied with railroads, television, and airlines.

My point is not that the Interstate Commerce Commission, the Civil Aeronautics Board, or the Federal Communications Commission do not deserve professional attention. However, academic literature and teaching need to take fuller account of the basic expansion in the scope and character of government regulation that has been occurring in the past decade. The expansion in regulation--whether measured by the size of regulatory budgets or by the number of rules--is in these newer areas. Whether the railroads and their unions "capture" the Interstate Commerce Commission is a far more trivial concern than understanding the full range of impacts of environmental, safety, credit, and employment regulation. That improved understanding is essential for developing support for reducing the many adverse side effects of regulation that we have been discussing--higher costs, loss of jobs, reduced productivity and capital formation, and a slower rate of innovation of new and better products.

Hopefully, academic research will help to shift the public debates on regulation to new and higher ground. It is not a question of being for or against a clean environment; of course, this nation is firmly committed to a sustained effort to reduce environmental pollution. But we need to turn our attention to the truly serious questions: Are we getting full benefits for the large costs that are being incurred to meet regulatory requirements? Are those regulations the least disruptive and the most-cost-effective way of attaining the nation's social objectives?

Far too often, the answer is that regulatory results fall short of what is desired. Economists have developed a simple but useful notion that I mentioned earlier: Overregulation is not an emotional term; it is simply our shorthand for regulations that cost more than the benefits they provide, that do more harm than good. It should not be surprising, therefore, that the change we most frequently urge in public policy is not to dismantle the regulatory apparatus but rather to introduce into regulatory deliberations concerns over benefits and costs. The benefits and costs of a regulation should be measured before it is issued, not afterwards. If that were done, regulators would have the opportunity to revise their rulings before issuance if estimated costs exceeded benefits.

This is an ambitious agenda for public and private action, but I have encountered no"quick fixes" that would cure all the shortcomings of the many efforts to regulate business. Perhaps recognition of that fact would set the stage for durable reforms of this aspect of business-government relations which has such vital impacts on the consumer. The need is not for dramatic confrontation but to improve the public's understanding of the massive extent to which regulations at every point of the building process generate costs that neither the government nor the public appreciates. Each of us must contribute to that educational process.

PRESENTATIONS

EDWARD L. SIMONS
Manager, Environmental Protection Operation
General Electric Company, Schenectady, New York

For centuries, society has been willing to tolerate pollution as a necessary by-product of useful technology. However, since 1952, a network of laws and regulations designed to reflect or mold public policy in the areas of environmental protection, safety, health, and energy has been developed.

Today, I want to focus on just two major developments: the Clean Air Act that was amended in 1977 to provide what is essentially a comprehensive federal policy on industrial development and the developing regulatory strategy in other areas of environmental control.

The Clean Air Act and the regulations being developed for its implementation represent a compendium of some of the most complex legal, social, scientific, and technical issues in the environmental area. I would be foolhardy to try to unravel them today; however, because of their importance, I will try to illustrate in an oversimplified way the basic regulatory strategy of the Clean Air Act (Figure 1).

For each of six major air pollutants, the Environmental Protection Agency (EPA) has established a national ambient air quality standard, which defines the maximum permissible level of that pollutant. Each county now has been identified in terms of attainment or non-attainment of the standard for each of these pollutants. In the clean areas, the pollution level is below the national ambient air quality standard; in the dirty areas, it is higher. By the end of 1978, each state was required to submit to the EPA a detailed plan showing how it would achieve, and subsequently maintain, the national ambient air quality standard in the dirty areas by 1983. If a state was willing to impose such politically unpalatable programs as transportation control plans, it could get an extension to 1988 for some of the pollutants.

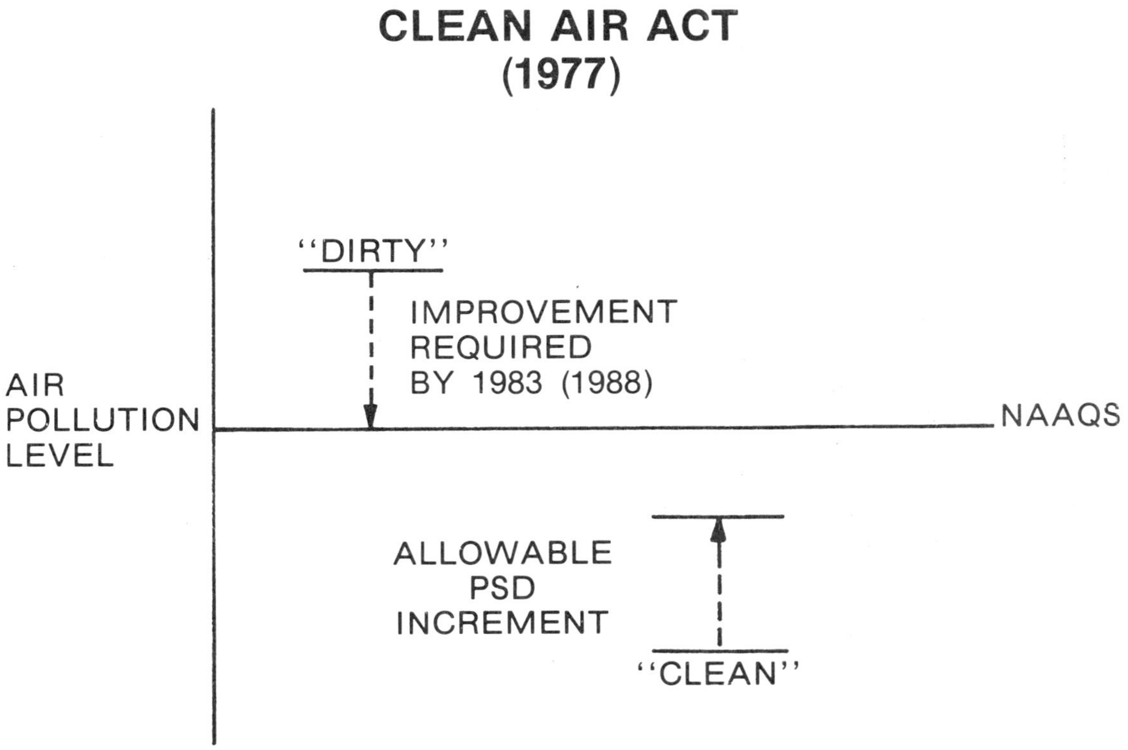

FIGURE 1 Clean Air Act (1977)

A discussion of the technology requirements that industry will have to meet is beyond the scope of this presentation. However, a key point for everyone who is involved in industrial planning for new construction is this: a state whose plan had not been approved by the EPA by July 1, 1979, was to be prohibited from permitting major new industrial construction or major modifications of existing plants after that date. As of July 1, only 35 states had even submitted their plans and only one, Wyoming, had been approved. For the present, the EPA is leaning over backward to avoid stopping construction and is continuing to process applications for permits, a procedure that takes three or four months at the very least. By the time this processing of permits has been completed, the EPA expects that it will have issued at least conditional approvals of most state plans. The key states to watch, those with critical deficiencies, are California, Michigan, New York, and Ohio.

In the meantime, industry has no detailed official guidelines as to how regulatory policies will affect plans for new plants or the operation of existing ones. All we do know is that more lead time and much more data will be required before regulatory approval can be obtained for any future plans.

Although the national ambient air quality standards define levels that must be reached by the dirty areas, they do not represent permissible levels to which air pollution can rise in the clean areas. Future increases in air pollution in the clean areas are limited to certain increments which, in the wisdom of Congress, have been deemed necessary to prevent what is called "significant deterioration of air quality." Again, the technological implications of this are beyond the scope of this presentation. Let me say only that the Clean Air Act of 1977 is, in many respects, a land-use act that has made air pollution considerations a major factor in the location of new plant sites.

For example, the recent selection by General Electric of a site for a new manufacturing plant, designed to go on stream in 1984, was the culmination of a search that began more than two years ago with visits by GE personnel to state environmental officials to assess the effect of expected air pollution regulations on industrial growth. This was the first time in GE's history that environmental considerations played such a key role, not only in the selection of the final site but also in the elimination of candidates that would have been acceptable in terms of all the other conventional site selection criteria.

Although the Clean Air Act may be the key environmental factor in determining where we can build new plants or expand existing ones, the question of how an industry operates its plants and even what it produces is determined more by the developing regulatory strategy in other ares. This new strategy represents a shift in concern from current short-term acute risks, such as fish kills and heavy smoke odors, to the potential long-term chronic risks, such as the carcinogenic, mutagenic, and teratogenic health effects of certain chemicals. In short, the emphasis is clearly on the control of hazardous and toxic substances. Although the Clean Air Act and the Clean Water Act are being used as instruments for such controls, the most potent weapons probably will be two laws that complete the cradle-to-grave coverage. Under the Toxic Substances Control Act, the EPA will be able to control the production and use of all chemical substances and, what may be more important to this group, under the Resource Conservation and Recovery Act, it will control the disposal of what it defines as hazardous wastes and, as you might expect, the definition is very broad.

The impact of these issues can best be illustrated by reviewing President Carter's August 1978 declaration that the Love Canal was a national disaster area. This was the first time in American history that such a designation had been applied to a man-made disaster. What and where is the Love Canal and how did it become a national disaster?

What we call the Love Canal today is a three-block area in the southern part of Niagara Falls. It is named for William T. Love, an entrepreneur who decided in 1894 that there would be an industrial future for this area if he could build a canal between the upper and lower Niagara Rivers. He went bankrupt, and during the early decades of the twentieth century, this segment of the proposed canal, which was all that had been completed, was used as a swimming hole. In 1942, it was purchased by the Hooker Chemical Company for use as a chemical waste disposal site. After all, the canal had been built with impervious clay walls to hold the water that it would contain. Dumping by Hooker ceased in about 1953 after about 22,000 tons of chemical waste had been disposed of there. Hooker covered the site with clay and considered it sealed. However, the area was experiencing a building boom in the early 1950s. People were moving into Niagara Falls and new schools had to be constructed. So, under protest by Hooker, the canal site was sold for $1 to the school board. Hooker included a clause in the deed that gave notice of its past use and that was intended to protect Hooker from future liability. That was a pious hope.

The property not required by the school board was sold to developers who built 100 homes directly adjacent to the former hazardous waste disposal site and 139 across the street from these homes. In 1976, after six years of abnormally heavy rains, a rising watertable resulted in the rusting of drums and the leaching of chemical wastes and subsequent percolation into the basements of the homes. The New York State Department of Health began to note startling health problems, including birth defects, miscarriages, and liver abnormalities. By August 1978, this was Love Canal.

After President Carter's declaration, the state began to evacuate families from their homes, homes that had by now become both dangerous and virtually valueless. By late fall, the creeping deterioration was more evident. The New York State Department of Environmental Conservation contracted with an outside firm, the NEWCO Company, and 86 days after the state had recommended evacuation, work was begun on excavation for a drainage system to channel away the toxic leechates. Expenditures of more than $25 million have been made to evacuate families, purchase their homes, and build this massive drainage system. EPA Administrator Douglas Costle estimates that the Love Canal site, built to the right specifications, might have cost about $4 million (1979 dollars) and spared the incalculable cost of human suffering.

Love Canal probably will end up in the lexicon as a generic term for chemical waste pollution. It has led to the public's cynicism over the existence of technology that is able to control hazardous waste disposal. The public just does not believe that we have the technology to build adequate sites and does not want them in its neighborhood. The message that we in industry are getting from the public is that nothing we use in our plants and nothing we discharge from our plants, whether as a finished product, an air emission, a wastewater discharge or a drum of waste, must represent a current or potential unreasonable risk to public health or the environment.

Although it is often popular to make American business the scapegoat for all of our environmental ills, this meeting and others like it show that most of U.S. industry does not want to ignore or passively accept the environmental impact of its operations. It is learning to conduct its business in an environmentally acceptable manner, and given a rational approach by regulatory officials, I am sure we can demonstrate how to do this in an economically sound manner.

CALVIN S. HAMILTON
Director of Planning
Department of City Planning, Los Angeles, California

Although my job is as a regulator, I think I can see both sides of the coin. I annually review the plans for developments worth many millions of dollars so I understand what private business faces in terms of costs and delays that come from different types of regulations. I concur that the impact of the environmental controls on air pollution probably will be more severe than that of any other land-use regulation promulgated in this country. I also happen to be the environmental quality officer for Los Angeles and in charge of approving all oil drilling, and since we have 2023 oil wells in Los Angeles, I am quite familiar with environmental problems.

On the other side of the coin, there are many horror stories from the public's point of view that cause government to institute tough regulations. In Los Angeles, 100 homes were lost in mud slides last year, all of them in subdivisions built before the grading controls were strengthened in the 1960s in response to a public outcry. We did not lose one house in those areas where we had instituted new strengthened grading requirements. Most of the laws we have on the books in this country that affect the cost of housing and building were not just dreamed up by bureaucrats. Generally, they are the result of substantial public pressure and neglect and diffused leadership on the part of the building community.

From the developer's point of view, there are a lot of things that we need to solve. The cost of governmental regulation is relatively low when compared to the current cost of money, but there are delays and duplication by different agencies. It is ridiculous that it took 900 different permits to build a pipeline from Long Beach to Texas. Overlapping reviews, multiple agency jurisdiction, the lack of certainty that results from changes in laws and the imposition of those changes during the processing of approval for a development, and the legal implications of environmental and other controls are serious and costly problems. The more regulations one must adhere to, the more apt one is to be taken to court. The interference in private decision-making by government is costly in terms of time, staff, and inflation.

One way to tackle the problem of governmental policy improvement is to improve the processing, reduce the number of agencies, and eliminate some of the regulations at the local level. Another way is to assess the scope of concerns and determine what most appropriately should be addressed by federal agencies, by state agencies, by regional agencies, and by local agencies.

Still another way is to determine public policy through functional concerns. For example, determine which concerns are logically environmental and how to deal with those collectively. One also could solve the problem of governmental regulation and its cost to private development by combining scope and function. With regard to planning, for example, we need a national policy on planning and urbanization that would be coordinated with other issues and would be dealt with at the national level. We also would need state and regional policies on urbanization and, finally, local planning policies that were compatible with the others.

The critical issue is to regulate only at the most logical level. What do I propose? I suggest that we need some type of national commission to define and identify our goals and objectives in the area of community development in this country by dealing only with national issues. The same should be done at the state, the regional, and the local levels. Goals and objectives then should be addressed by functional concerns--that is, what are our true concerns in the environmental arena, in the planning arena, in the economic arena, and in the physical development arena. Present legislation at each jurisdictional level then should be analyzed in relation to those preliminary goals and objectives. It is necessary to determine who is regulating whom and why and how it could be done better to accomplish our goals and objectives. This should not be too difficult since there are literally hundreds of existing studies to draw upon. Finally, the various legislative entities at the federal, state, regional, and local levels must be stimulated to allocate responsibility to the logical level of government to consolidate responsibilities, to avoid duplication, to simplify procedures, to reduce the time involved, to place responsibility at the most grass roots level that is possible, to have a "sunset" review procedure at each level of government, and to crosscheck the legislative proposals with the basic goals and objectives at their functional and jurisdictional levels.

You might ask whether this is a pipe dream. Well, it probably is. However, if this industry organized to support such an approach, it might find it has more muscle and resources than most other industries and businesses. I truly believe that if you look at needed government regulatory policies in a logical fashion, you can accomplish massive changes.

JOHN W. GUINEE, JR.
Real Estate Investor
Reston, Virginia

I submit that the little things, the low level activities of counties and municipalities, are as bad or worse for the industry as the higher level federal problems and that they can be attacked and solved much faster than anything at the national level. One of the earlier speakers referred to the building industry as a cottage industry. Well, it is difficult to fix through federal action a cottage industry once it goes wrong. I will primarily address inspection procedures and what I think can be done to them because they represent a problem that spans the length and breadth of the land.

The question as to whether or not regulation will reach the point at which the costs justify the added benefits is, I think, rather academic. No one knows whether we are beyond that point or not because no one has done any work on determining the cost-benefit relationships of these bits of new regulation. "Bits," I think, is the operative word here. Most regulatory revisions that have come into existence over the past several years have been little ones or perhaps several little ones codified and adopted simultaneously. None of them has been important enough to get anybody upset enough to prevent them from coming into existence. It is only when the sum of the whole thing is looked at that the little bits become important.

When I first became involved in the home building business in the Washington area about 20 years ago, five inspections were required to build a house. They were time-honored, traditional, and accepted. Now there are many more--13 in fact. They are not so time-honored, traditional, or accepted. But I am afraid, however, that eventually they will be accepted and the cost to the home owner will be an accepted evil--accepted but not acceptable. I do not think all of these inspections are justifible and I believe now is the time to stop them.

Consider this example. Most houses in this area have basements. A prerequisite to a basement is a hole in the ground; you then build a suitably waterproofed basement, backfill it, and that is it. There was a time when backfilling was not inspected and, therefore, the foundation was not inspected. But then there was a rash of wet basements one spring following one bad winter, and suddenly a new inspection came into existence, a 100 percent inspection to solve about a 3 percent problem. At this stage of construction, each inspection costs from $10 to $20, plus the fee since one day must be taken out of the construction schedule for the inspection. That $10 to $20 for a $100,000 house may not seem significant, but it is since there are 13 inspections, and at $10 to $20 each their cost would pay for a dishwasher. Although one a day is allowed for the inspection, it is rarely made in one day, particularly in the height of the building season when you need it most often. In this past season, it took up to 13 days in Fairfax County, Virginia, to have one inspection made after it was requested. We had about 500 houses in the pipeline this year so that 6500 inspections were involved. At from $10 to $40 per inspection--the $40 allowing for only a 1 day delay per inspection--the cost to the company was between $65,000 and $250,000. That is about $500 per house. The customer, of course, paid for that. We just passed it on.

But what is going to be done about it? Three things. First, there will be suits to get what might be called a continuing mandamus to require that the inspections be held on the day following the day that they are requested. I do not yet know what the penalty will be if they are not. It will probably take two to three years to get through the court cycle in Virginia, but at least that effort will be started. Second, there will be drafted, and pre-filed for the next session of the legislature, an amendment to the state law. This amendment also will require that inspections be conducted by the end of business on the day following the day of request. It might require two, three, or four sessions, but eventually that law will be put into effect. It is getting too expensive for anybody to ignore the time loss factor and the resulting impact on productivity.

These are both stick actions. The third action is a carrot. Taking a page from the book of that fine, efficient institution, the U.S. Postal Service, some in the industry have suggested that, until the lawsuit is won or the law is changed, any builder who wants next-day service should be able to pay an extra $10 and get same-day or next-day service instead of four-or-five-day-later service. It will cost a little bit more up front but it is worth it. Have you ever tried to explain to 30, 40, or 50 families that their houses are 2 or 3 weeks late and they cannot move in before school opens because the county did not make its inspections on time? Believe me, that little bit more money will be worth it!

I also see no need for a 100 percent inspection on anything. Statistical sampling techniques should be used for building inspections just as they are for other things. Fairfax County has been trying to figure out how to do it for three years, but it does not even have a data base yet. There is no particular reason why every builder has to go through every inspection, all 13 of them, on every house. Building departments know who the good builders and the bad builders are and where the wet ground and the dry ground and the bad ground and the good ground is. They should be able to make a 5 or 6 percent inspection and then let the builder go ahead with his work. I think it will take some time to make this change, but it is necessary.

Finally, I too think we need a "sunset law" on every single regulation that is established. Every new inspection or regulation should have a termination date. If it is a good regulation, there will be no problem in re-installing it, extending its life. If it is a bad regulation and cannot stand the test of time, it might as well die.

In addition, many inspections are redundant. Many manufactured items that are installed have already been certified by Underwriters' Laboratories, Inc., or a similar organization, and there is no need to reinspect them all again.

If we are to improve productivity, we must modernize. We must make the regulations die in a timely fashion. We must eliminate the completely nonproductive days when nothing happens while we wait for an enforcement agency to do something.

CHARLES J. DINEZIO
Executive Director
State Building Code Commission, Boston, Massachusetts

The building code itself is not the log jam in the permitting system. There are many other things that stop that process. Many participants in this Forum have discussed overregulation and I agree with them. But I also think regulation has helped.

Prior to 1968, about 15,000 municipalities in this country had independent local building codes. That number has been reduced substantially since the Douglas Commission report was published in 1968. Since 1970, all-encompassing legislation has been enacted by many states (first by Connecticut and, second, by Massachusetts) that creates a state building code commission empowered to regulate the entire building regulatory system as we understand it. This system includes the development and promulgation of a statewide building code which cannot be minimized or maximized. The designer can, but the local municipality cannot.

In Massachusetts, there were 351 different building codes--351 different fiefdoms and opinions--before the state building code became effective in 1975. Few builders, architects, or engineers would want to revert back to what we call in New England the "home rule" syndrome. I remember one community in which you could build only masonry structures and other places that permitted only one type of wood framing while neighboring areas required a different type. Thus, regulation has vastly improved the system of building construction in this country.

In Massachusetts, the initial recommendation made to the general court was that, in addition to a statewide building code, uniformity of administration and enforcement are vital to a uniform building code. This is still severely lacking. One, obviously, cannot expect a building code to be administered and enforced by an untrained person. Professionalism is required.

How was the enabling legislation that established the State Building Code Commission in Massachusetts enacted? Through the collective effort of the builders, contractors, engineers, architects, League of Cities and Towns and League of Women Voters. Surprisingly, of 300 people attending the final public hearing, not one opposed the legislation that would abolish home rule. More than 50 percent of the local building departments also agreed that one state building code and one set of regulations were needed. It may be that the time has come to use such a cooperative effort to develop one national building code.

Massachusetts has been a leader in a variety of areas and particularly in the building regulatory system; yet I feel that I direct one of the smallest state agencies in the United States. We were one of the first states to promulgate rules and regulations for laboratory accreditation; yet we were unable to obtain the necessary development funds. We were, however, fortunate to obtain a $40,000 grant from the National Bureau of Standards in 1974, to develop a laboratory accreditation program for freshly made concrete and also a system for licensing the individuals who do that testing. This program was established as a direct result of a 16-story building collapse where four workmen were killed. We were crisis-oriented, so the legislators had immediately filed legislation in response to this catastrophe. The Massachusetts program is now leading toward a national voluntary laboratory accreditation program. This national program will certainly benefit the construction industry since it will be guided by a single set of rules and regulations.

Crises should not stimulate the promulgation of regulations, because we then tend to overregulate. Think about the myriad of regulations issued following the Coconut Grove fire!

Our life loss record with high-rise buildings today is fantastic--all due to our improved technology. Uniformity of building codes, I think, also has added another dimension.

But what do we do when we leave meetings like this? Do we simply go back to our cocoons and talk about the lovely discussion we had? Hopefully, the construction industry will stop talking about divisiveness. Hopefully, the trend for the 1980s will be cooperativeness.

You have the means to effect change. You deal with the politicians on a day-to-day basis. I can do very little, since I am an appointed individual in state government. You, however, have the "muscle" and, collectively, you can make things work. State agencies are obliged to go through a public hearing process. If you do not like the regulations proposed, change them. I have to deal with the special interest groups, and that is very difficult politically. However, if you can do something in one state, you can stimulate a lot of other change. After all, it only took 8 years for 17 states to adopt mandatory building codes.

I frequently feel that my five-year plan for regulatory reform actually is a ten-year plan. It is very frustrating to work for government, to think that what I am doing now I should have been able to do in 1975. It is up to you and your organizations to use the elective process to indicate that funding and qualified professional people are required to deal with the regulatory process. We need education for building officials. We need more professionalism. This is a very slow process.

Massachusetts has mandates to develop and promulgate a certification program for building officials and a licensing program for construction supervisors. We have a laboratory accreditation program and a products approval program. However, I am not able to develop the former and to implement the latter programs effectively because of lack of funding.

To digress for a moment, let me tell you about a contractor who called me recently about a problem he had encountered contructing a Class A building with apartments and a garage below. His original submission included a reinforced concrete curtain wall between columns. He decided that he could save $22,000 by replacing the curtain wall with concrete blocks. When he told the building official what he wanted to do, the official said he could not do it and must follow the plans as originally approved. When a building official makes such a decision, he must cite the section of the building code covering the specific requirement. If he cannot, you are not obliged to do it. In this case, the contractor called me. I looked up the appropriate section in the building code which allowed the desired change. He decided not to proceed because of his concern over possible future actions the building official would take during construction. You cannot blame the regulatory agency for that contractor's decision. The contractor on the job must make his own decision.

I also testified before Senator Proxmire's committee when it was dealing with rehabilitation and noted that he devoted some time to a discourse on corruption in building departments. I would just like to note that it takes two people to corrupt, the giver and the receiver.

With respect to rehabilitation, let me give you an example of the cooperativeness that can take place. In 1978, Massachusetts was focusing on the revitalization of its cities and towns. The State Building Code Commission was requested to develop building code provisions to deal with existing buildings even though there were no guidelines to follow. Again, there were no development funds for this purpose. Consequently, the governor appropriated $48,000 from his discretionary fund, and I was able to obtain help from the National Bureau of Standards. We then got the three national model code groups, the National Conference of States on Building Codes and Standards, the Association of Major Building Officials, of which Creighton Lederer is a member, the National Association of Housing and Redevelopment Officials, and other groups to contribute and ended up with over $200,000.

We probably should have included the construction industry (architects, engineers, the developers), the insurance companies--because of the direct benefits they would gain and because of their interest in the revitalization of our cities and towns.

The U.S. Department of Housing and Urban Development contracted with the National Institute of Building Sciences to have three guidelines developed. These guidelines were printed in the Federal Register this week, and I ask that you, who are going to be affected by them, look at them carefully and comment on them. What I am getting at is that we cannot solve the problems of the nation from the local or the state level, but that we can collectively contribute to their solution.

One thing that I would like to see emphasized by this meeting is the need for a better understanding of those who are being regulated and those who produce the regulations. More importantly, I think that we can collectively express to our elected officials that we have a problem in this country--that it is one thing to enact legislation to establish building regulatory systems for the benefit of all those affected by the system, and another thing to develop and implement them without adequate resources.

CREIGHTON C. LEDERER
Director and Commissioner
Buildings and Safety Engineering Department, Detroit, Mighigan

Many of us are aware that, in the regulatory area, loss of time in the permit process costs money and is an indication of a regulatory problem. Many cities have gotten into trouble because they did not have a smooth-flowing permit process. The permit process can be onerous. It is not understood by most people in the building industry, and it generally is perceived as a barrier.

How do we speed up the process? How do we make it less painful? One method we explored in Detroit was a "one-stop service" developed to reduce plan review time and to enhance the understanding of codes by the people applying for permits. In establishing our system, we investigated three types of "one-stop service":

Plan 1--having all approvals made in one location.

Plan 2--having a coordinated and monitored system for routing plans to the responsible agencies.

Plan 3--having a central authority as well as a system for routing plans on a controlled and monitored basis to the responsible agencies when necessary.

Plan 1, which involved having all the controls centrally available in one place, is not practical except in the smallest cities and the simplest cases because of the number of approvals required. A desirable aspect of Plan 2, having the plan review done in various agencies, is the checks and balances that qualified code experts provide. In the real world, however, these experts must be used efficiently which means they have other duties that do not lend themselves to working in a central location. Most of our plan examiners double as supervisors, code committee convenors, and training staff. Since we wanted to supply a fast, efficient service by qualified people (all our plan examiners are registered professional engineers, which addresses the problem of effectively controlling the process with people who have the ability to do the job), we decided to use Plan 3, which involves central control and monitored plan routing. It depends on the applicant supplying adequate and honest information. Anyone who approaches a regulatory body and does not lay out what he wants to do in an honest and forthright manner is bound to have trouble. In Detroit recently, someone requested a permit for a small manufacturing building located close to the central city. The plan submitted did not disclose, but we found out, that one of the products of manufacture was phosgene gas. This was just not the thing to try and run by us without disclosing all the aspects--good and bad--of the project.

To be effective, "one-stop service" needs standardization of codes and of the process for getting a building permit. Mr. Dinezio talked about the thrust toward model codes in the United States. We have a model state code in Michigan. Detroit is the largest city in the United States that has given up its own big city code and uses the state code. This cooperation between the state and the city is one of the key aspects of making the "one-stop service" process work. Any builder coming into Detroit knows the basic requirements that must be met. The building code is not a problem.

How does speeding up the process actually work? We have found that the pre-application conference at which you identify the problems in the building and sort out the need for intepretation is invaluable. It is the secret of how to make the "one-stop service" process successful. Does the pre-application conference work for rehabilitation? The designer, engineer, or architect very quickly becomes aware of the pitfalls of making a building code developed for new construction applicable to such work, but much of the housing work in major cities throughout the United States today is rehabilitation. We in the building industry have not adequately addressed the adaptive re-use of our buildings on a national scale. Thus, the pre-application conference is very important because it puts the problem of codes and rehabilitation on the table. This gives the builder the opportunity to find out what the regulatory process demands. It also shows where there is a need for any code interpretations or minor variations or where major variations in the code are needed.

We use two levels of review. One is an administrative committee that meets weekly. It has the authority to recommend action on interpretations or minor variations to the code. Most minor problems are, therefore, solvable within a week.

For major problems (e.g., those involving a design that does not meet the code or the adaptive re-use of buildings where the code is regressive), we use a Board of Rules that has the authority to consider major modifications. This Board, a nine-member committee consisting of representative from industry, city government and the public, has the authority to consider and to act on major modifications or waivers to the code. Because of the successful operation of the Board of Rules, the authority for addressing permit problems centralizes itself in the Building Department. We also are responsible for implementation of the housing code, the zoning ordinance, the energy code, and the barrier-free-design ordinances. All are taken care of in a central location. With this process, we try to provide the developer or the builder with everything he needs to efficiently obtain a building permit.

Our "one-stop service" is based on good information flow, on good communication, and on understanding. The building official is there to serve the public. He is there to protect the public. He also is there to protect himself. The builder has different goals. He is serving a client. He has cost in mind, which is very important. It is necessary to be aware of the differences between the two functions. They often are conflicting.

We also have used the "sunset law." It is an excellent idea. It works on the most controversial ideas. Our All Sales Inspection Program, for instance, was given a life of one year after vociferous public debate. At the end of the year, the effect of the law was reviewed and it was adopted permanently.

Processes such as "one-stop service" and "sunset laws" work. They are available to us. All we have to do is to cooperate and understand to make them work.

JAMES M. BROWN
Professor of Law, National Law Center
The George Washington University, Washington, D.C.

My remarks will be focused primarily on a matrix illustrating the relationship between regulations and codes and the building process (Figure 1) and on some of the potentials suggested by what it portrays. Let me preface my remarks with a few thoughts on some educational needs and potentials.

One element of the educational process that has been largely overlooked by the building industry, among others, is the prospect and need for exercising initiatives to ensure that universities around the country have the information and the capability to impart to their students a sound and comprehensive understanding of the complexities, capabilities, responsibilities, limitations, and constraints that are associated with the building industry today; an understanding of similar complexities existing within and between the various echelons

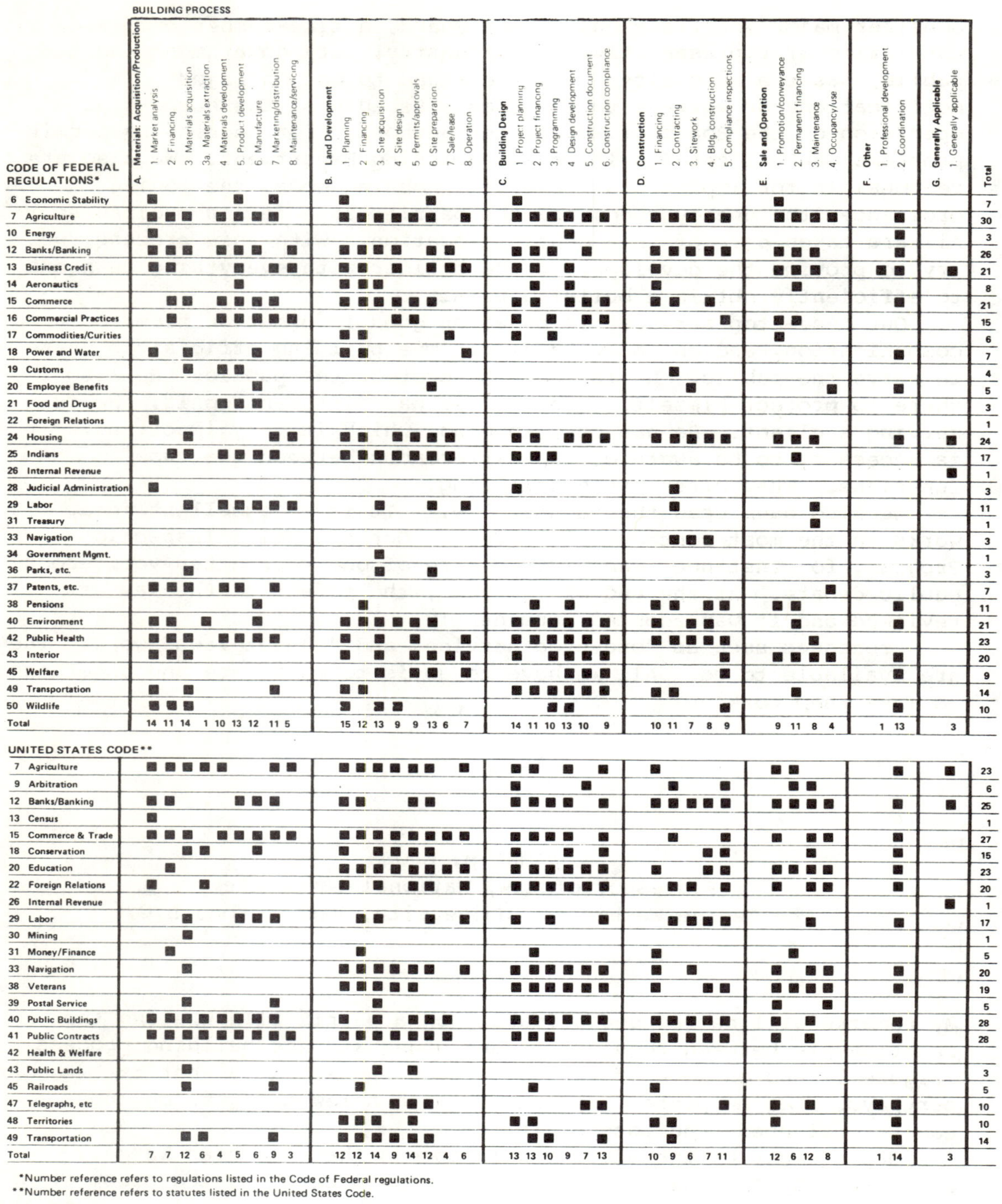

*Number reference refers to regulations listed in the Code of Federal regulations.

**Number reference refers to statutes listed in the United States Code.

FIGURE 1 Regulations and codes affecting the building process (from National Institute of Building Sciences, A Study of Regulations and Codes Impacting the Building Process, Washington, D.C., 1979.

of government; and an understanding of the complexities and, often, the diseconomies attendant to the myriad points of interaction between the two. No matter how dedicated various university departments and faculties are to meeting their responsibilities to their students, they cannot keep abreast of the current situation without continuing, constructive interactions that permit and stimulate the development of knowledge about current problems. I suggest that the traditional alternative is, of necessity, largely an historical one, interesting and essential for learning of past successes, accomplishments, frustrations, and failures but not always dependable as to the real reasons for what occurred or, by itself, the best training for developing a ready capacity to cope with new problems.

The young U.S. population is largely concentrated in urban-suburban areas. These students generally go to school for a longer time than did the preceding generation. They do not have the kinds of quickly maturing exposures and demands that you and I had when we were growing up. These young people have been delayed in their development and in their exposures to what the real world is all about. They are idealistic, intelligent, and, in many ways, well educated, but they have been considerably slowed in incurring a comprehensive cross-sectional exposure to an "uncloistered" life. As they work up through the academic process, they absorb progressively more filtered summary data that are built on events, consequences, and condensed analyses of cause and effect. This is true, perhaps, more in the humanities areas than in the hard sciences, but the social side of the hard sciences suffers even more from a condensation of socially significant aspects of professional activities. As a result, at the professional school level, the degree of naivete of many of these well meaning, concerned, and interested students is disturbing.

One of the problems I experience in trying to teach in my particular area of responsibility is to offset misconceptions and the tendency to deal with all problems on a "national policy" level. I have concluded (partly on the basis of about 15 years in the construction, land development, and building material businesses before I went on to a career in law and law teaching) that one cannot effectively teach what needs to be taught about such common things as zoning and building codes and subdivision restrictions and exactions, much less about all of the proliferating areas of environmental, natural resources and land-use management and control, through a pure textbook method. We may be able to impart theory and, for what it is worth, a few legal principles and their application by the appellate courts, but, using the traditional approaches, we cannot impart any real understanding of what is happening and why and with what consequences in the day-to-day operational and managerial decision arenas.

Therefore, I have been experimenting under the premise that to impart a necessary level of understanding requires a considerable amount of actual and/or vicarious exposure to "the real thing," with an attempt to achieve appropriate exposures by the use of accelerated-time simulation processes--a "gaming" of real life, real-time situations, wherein the students are faced with having to deal with today's needs and details under simulated real-life responsibilities and objectives of the day-to-day variety. This is a completely different approach, especially in the law schools, from traditional education. The demands on teacher and student are much greater but so are the rewards in understanding, accomplishment, and functional preparation. One major problem with this approach is to get sufficient real-life detail to ensure a realistic simulation. This requires a tremendous amount of the "impedimentia" of the day-to-day existence burdening the typical building industry decision maker, public or private. To hold a zoning hearing, for example, I need the same type and amount of supporting material that you would have. To get a building permit or to get a subdivision plat approved, I have to provide the same things and to build in the same frustrations that you experience, even down to trying to ascertain just what clearances and approvals and acceptances are required and from whom. "Gaming" has to "shadow" real life and must provide the same institutions, tools, and representative personalities, the myriad alternative choices, the same degree of variables, etc., as exist in fact; if it does not, it is not adequately representative. We can compromise on quantity but not on detail. The wherewithal to do this is not found within the covers of textbooks; it can be assembled only by "scrounging" for the necessary "impedimentia" from tolerant and cooperative people out in the field who appreciate the thrust of our efforts.

It would be tremendously beneficial to all parties if the building industry actively involved itself in helping to initiate and to implement the capacity for educational institutions to accurately convey to interested students an understanding and appreciation of the real story of what goes on "out in the pits" where you are struggling. They do not understand it today. They are idealistic, and they deserve to be commended for that. They are receptive to honest information, are far more open-minded than they often are given credit for, and are aware of, and even a little frightened of, their lack of exposure to reality. They are concerned about whether they can "cut it on the outside," and they want to be able to earn their way from the start. They strongly sense that it takes more than educational theory to be able to do that. They will appreciate sincerely your interest and involvement in bettering their education.

If you do elect to try to help, please be aware of one basic fact: These students may be naive, but they are not pushovers for a "con job." They will be very quick and very sharp to discern a "propaganda pitch" if one is thrown at them, and whoever tries that will destroy his credibility. They will be particularly skeptical if the "con" is pro-industry or forthcoming from industry. But you do not need to con them. You need to earn their respect, and you can if you level with them, even though they may be inclined to disagree with your stacking of values because you, as an industry, merit respect for the remarkable history of accomplishment you have created. If, along with such earned respect, you also receive some sound, constructive, knowledgeable criticism instead of childish platitudes, would we all not be better off? After all, they are our kids--and it seems doubly foolish to misrepresent what makes us tick just before we turn the business over to them. Who is going to be in a better position to call us out, and who should we want less to cripple with misinformation?

Having made that pitch, let me tell you a little bit about the history of the matrix (Figure 1) and suggest how and why the information it portrays has inherent potential for improving productivity in building and construction. Keep in mind that this focus, too, is on educational potential.

Slightly over a year ago I was involved, in a consultant capacity, in an effort, contracted for by the National Institute of Building Sciences (NIBS), aimed at identifying those segments of the U.S. Code and of the Code of Federal Regulations (CFR) that exercise a significant economic influence on the building industry of the United States. It quickly became apparent that there is no dependable source where we can find out easily and simply what federal statutes and regulations have an impact upon the construction industry. Even worse, we have no process through which government officials are informed easily and dependably of the comprehensive effects of their determinations on the industry. We have no process or institution presently functioning to aggregate and provide such information on an industry-wide basis to those who might actively seek it or to disseminate it selectively to those who should desire it as an informational prerequisite to their decision- and rule-making.

To conduct the study referred to, it proved necessary first to develop a detailed description of the construction industry and its functional components in a format that allowed a cross-matching with the federal codes and regulations. (We did not attempt, in this effort, to undertake the compounding effort of probing state and local levels). It then was necessary to devise a process for identifying, out of the massive bulk of primary source material, the titles, chapters, subchapters, parts, subparts, sections, subsections, etc., that were pertinent and for selecting what, of that material, was significant.

The sheer bulk of such material is awesome and is rapidly growing. For example, if stacked into one pile, the CFR would extend approximately 18 feet into the air (compared with about 3 feet in 1938 and about 4 feet in 1958). The U.S. Code Service volumes, the most effective code set for the purpose, would make a stack about 19 feet high. If one decided to work with the U.S. Code Annotated, the stack would be 33 feet high. If one plowed through the Federal Register for background, the volumes printed just since 1970 would stack 90 feet high.

A considerable selection, screening, and summarization effort accompanied the extraction of raw data, which were collected, given coded abbreviations, and entered onto work sheets. From the worksheet, summary pages were developed to the extent that, when assembled for a workshop, they covered one wall of the room. On those matrices were the coded abbreviations of the code and regulatory citations and titles that we had selected as affecting the industry in a significant way. Those multiple entries in each matrix "box" then were represented in a gross summation format, which is Figure 1. Each black square thus indicates at least one, and usually several, codes or regulations that are significant at the particular interaction point designated.

If this potential for data accumulation and organization was expanded to its necessary level of detail, it still would be merely a first step by which one could identify each federal influence and determine what segments of the industry were probably impacted. A logical next step, which we only tested by means of a workshop activity, would be to ascertain for those in the field, which of the many identified points of interaction were functionally significant, in what ways, to what degree, and with what attitudinal reactions.

Such an analysis could provide a basis for determining just what segments of the federal process were most needful of attention; reasons for such conclusions; and suggestions as to the type, level, and degree of response needed and as to where it should be directed. Even this capacity would be far from adequate.

Still needed would be an appropriately trained institutional staff, perhaps within the Building Research Advisory Board or NIBS, with the capacity and responsibility to intercept recommended statutes and regulations, orders, directives, etc., at the study, discussion, and evaluative stages; to analyze them for their potential effects on the building industry and, through it, on the nation; to carry that analysis, and any recommendations stimulated thereby, back to the official decision process while there was still opportunity and time to react in a reasonably appropriate manner to the information offered.

I am not advocating a lobbying effort in the typical sense of that term. I am suggesting an educational process, a process for identifying or developing the levels and types of information that all responsible decision-makers should desire and that all too often is not available. This should be a two-way process--carrying information of contemplated prescriptive or regulatory offerings down to the people in the industry who will be affected by the day-to-day applications as well as predicting the probable consequences and passing information and assessments thus obtained back up the line for appropriate consideration. The same process would serve to provide continuing post-decisional feedback so that the legislative process could function, in the dynamic sense, more closely in tune with its purpose and responsibility.

If the building and construction industry, as the largest industry within our economy and as the provider of basic essentials of existence, is unnecessarily, improperly, dysfunctionally impeded in the pursuit of its basic purposes and responsibilities by the proliferating interactions with government, it can and has a duty to mitigate such detrimental constraints and impediments. To do so, it must inform and educate itself and develop a capacity to pass its observations and knowledge on to appropriate recipients. The small effort I was privileged to participate in demonstrated an existing inadequacy and a need to conduct a comprehensive and detailed survey, in the nature of our experiment, with the results made available for self-education and for the constructive, objective enlightenment of governmental officials. Without such an effort, continuously maintained by an appropriate industry representative, we can expect no slackening in the uncoordinated proliferation of regulations requiring compliance and often impeding the smooth functioning which is essential to a healthy construction industry.

DISCUSSION

QUESTION: Does not the federal regulatory process reward adversary action quite independently of the search for reasonable standards?

DR. SIMONS: I can give one comment on an area that is not directly related to the building industry but has come up in the environmental area. The Environmental Protection Agency (EPA) has been developing a massive body of regulations under the Clean Water Act, and when it issued regulations in June 1978 that established a system of fines for the spills of hazardous substances, the Chemical Manufacturers Association (CMA) decided that the regulations had been written without proper regard for what the law required. The CMA initiated suit in federal court, which resulted in those regulations being thrown out. That was a clear adversary relationship. The CMA sued and the EPA lost. Congress amended that section of the law and then the EPA issued a better set of regulations. This year the EPA again issued a set of regulations under a different section of the law, and the CMA felt these were quite deficient in at least 45 particulars. This time, the mere threat that the CMA would take litigation was enough to get the EPA to sit down and talk. As a result of that negotiating approach, 43 of the 45 issues were settled to the mutual satisfaction of the parties involved. Thus, the key seems to be that the regulated party should understand precisely the effects of the proposed regulations and, better still, the alternatives that are available. The regulatory agency has a legal obligation to propose regulations, and it is not enough to simply state that you do not like the regulations. You also must explain what can be done instead.

MR. GARDNER: I would like to expand on that since Dr. Simons and I work side by side. He and our attorneys very closely monitor proposed legislation and do quite a bit of work with the legislative bodies in making recommendations on ways to implement new legislation. Dr. Simons is very instrumental in making better laws.

MR. LEDERER: Those of us on the enforcing end of the regulatory process are sometimes amazed that the building industry, with all its clout, does not make the effort it should when the rules are being made. I watched this happen in Michigan when the state building code was adopted. Architects, engineers, and builders had relatively little input until after the law was established.

MR. HAMILTON: My staff develops new codes, and we try very hard to involve the key interest groups early in the development process. We invite these groups to preliminary task force meetings and then send early drafts to a list of from 500 to 1000 different people. By the time we finish this process, we generally have pretty good support since we have developed the regulations with the particular groups that are going to be affected.

QUESTION: Mr. Guinee, you mentioned inspection delays. Did you mean to imply that inspection guarantees that you will not get a wet cellar? How good is inspection and how much does it truly ensure?

MR. GUINEE: You know the answer. Inspection does not guarantee anything. With respect to the inspection of basements, the only thing that could be seen prior to backfilling would be a very obvious skipping of parging or waterproofing whereas most wet basements result from things that take place during the backfilling process, after the inspection, or from hydrostatic pressure, which cannot be "inspected." In addition, during the electrical, mechanical and plumbing inspections, the inspector really does not inspect. The inspector will make sure that the wiring does not pull loose from the junction box or that a commode does not rock, but he adds very little to the project in terms of guaranteeing that quality and workmanship are adequate.

MR. DINEZIO: Mr. Guinee's view presumes that all people are honorable and do the right thing. If that were the case, we would not need policemen. However, there are a lot of bad builders who will, for example, put only one nail in a shingle or one nail in a stud. Who is going to check that? I think that the lending institutions and the insurance companies have an obligation, too. They take money from depositors and, hopefully, lend it out in the same communities. There are provisions and clauses in those mortgages that, for example, permit foreclosure if the property is not maintained. The insurance companies also have an axe to grind, but since the loss ratio is not very great, they ignore the problem. If these organizations would act, the building official might be relieved of some inspections. I agree with Mr. Guinee that the building official will not make every inspection if he tends to think that a builder is good. In fact, it is physically impossible in the United States to make every inspection. I know of no building department in Massachusetts that is sufficiently staffed to make every inspection required by law. However, the liability factor is becoming more important. The building official is becoming liable. If he does not make an inspection, is someone going to sue him? Will the municipality protect him?

MR. HAMILTON: I agree with Mr. Dinezio. Building officials do get sued. If a builder does not provide proper drainage or adequate compaction, the building department is in trouble if it does not inspect because the developer usually is gone before the problems occur. Thus, public officials often adopt regulations just to defend themselves. I regularly have about 34 lawsuits against me. When I came to Los Angeles, one lawyer in the city attorney's office was dealing with planning and zoning and environment. Today there are 15 full-time city attorneys doing nothing but defending me in lawsuits.

MR. GARDNER: The individual having a house built by a one- or two-house builder is protected by codes. Such a builder may not have the knowledge to understand why a building has to be built a certain way; therefore, the codes protect the consumer.

MR. GUINEE: I do not think any codes should be discontinued, but I was trying to make three points. First, anything that comes into existence ought to have an automatic death sentence under a "sunset law" provision. Second, statistical sampling techniques ought to be used instead of 100 percent sampling to make sure that those codes and those regulations which do survive are enforced in a logical, as opposed to a very expensive, manner. Third, the municipality has an obligation to do what must be done in a timely fashion. I do not care how many codes there are if they are justified, but I do care if they are never reviewed for effectiveness or if the required inspections are not made on time because that increases cost and decreases productivity.

QUESTION: Mr. Dinezio, how do you define your concept of one national cooperative code and how would it be developed? Would the federal government be involved in implementation and enforcement?

MR. DINEZIO: With respect to the latter, I would hope not. There are, as you know, four codes in the United States; three of them are model codes and one is developed by the insurance companies. Recently, the three model code groups announced they were getting together to do some consolidating, but they now apparently are rethinking that particular issue. Three organizations exist, and, supposedly, they are telling us that competition is good. If a manufacturer cannot get something approved by one group, he can go to another. That is just a lot of horsetrading. One document is sufficient. Why should a builder have to build to 351 different codes in Massachusetts? He does not have to do that in Massachusetts anymore so why should he have to build to four codes in the United States? Although the three model codes supposedly are 75 to 80 percent the same, the formats are different, and a builder very familiar with the code in one region will not understand the code in another region.

QUESTION: Is it reasonable to have a uniform national code in view of the fact that the climate in the 50 states varies from tropic to arctic, desert to rain forest?

MR. DINEZIO: Absolutely. Provisions in the technical sections of the codes already deal with that. Consider, for example, that the Canadians have done a far better job with codes than we have. Although the Canadian form of government might have something to do with it, the fact remains that their research is better. I think we can utilize federal agencies such as the National Bureau of Standards to do a lot of basic research for us. We can use the assistance of the federal government in that way and then direct the results to the state governments for implementation.

MR. LEDERER: Good regulations should be amenable to change. This is one of the problems of having the federal government controlling a national code. A good national code would require a consensus whereby the code could be amended. Otherwise, the situation would be worse than it would if we had 4 or 4000 model codes.

DR. SIMONS: Although the EPA may issue nationally applicable regulations, all of the major federal environmental legislation permits a state to promulgate its own regulations, provided they are not less stringent than the federal requirements. Is there any legislative impediment of that nature in the building codes?

MR. DINEZIO: You are obviously talking about the statutory authority and the power that state government has to rule. The states already have gotten together in only seven or eight years. During the next decade, this cooperativeness will become stronger, and there will be only one code in this country. The changes and modifications in codes made in response to special interest groups are political in nature. For example, wood-burning stoves have been a problem in Massachusetts this year. There were no standards and no laboratories accredited to do any testing. We now have these. Somebody bought $60,000 of unlabelled pipe, and he was able to get the legislation changed so that he could sell it. That is the type of political influence I am talking about. I do not disagree with the "sunset law" idea, but I think we must be careful in our evaluation.

MR. HAMILTON: The idea of a national building code makes me a little nervous. If we had one model code, the code group could institute changes and demand that every local agency either carry them out or provide an alternative. The EPA did that to Los Angeles when it told us to eliminate 80 percent of all automobile traffic. Our whole economy would have collapsed, but the EPA was absolutely serious. I suggest instead that we decide what makes sense at different levels and provide guidelines at different levels. Then we would have some flexibility. Some European countries have national building codes and some states have a statewide code, but these places are relatively small. I get very nervous about doing it for the nation as a whole.

MR. HACK: I have a comment on the question about adversary relationships. I have observed in my 26 years with government regulatory agencies that the responsibility for the regulatory function has shifted from the technically knowledgeable professionals to the lawyers. The lawyer in our society is trained to live in an adversary system (i.e., there are two advocates and someone judges), and although that system may be very good for adjudication, it is now the trend in federal government. The creation of the Department of Energy (DOE) is a classic case. The General Counsel came from the Federal Energy Administration (FEA) and decided that everything was going to be done by regulation, even operating procedures. That decision may have been a major reason for DOE's slowness. I wonder if the same thing is true on the local level? I wonder if Mr. Brown could tell us whether something is being done in the schools to teach those attorneys who will be government regulators that there is a substantive issue involved so that regulation does not become an end in itself?

DR. BROWN: Not very much is being done, but there is a gradually emerging interest, a gradually emerging awareness, among the law school faculties around the country that there is a segment of this thing we have not defined. Many law students have this awareness. Once one goes into a teaching mode, one is removed, at least one step, from the real world. But how does one maintain familiarity with that real world? As one's field becomes more and more esoteric, one tends to lose more and more contact with reality. That is part of the danger of the teaching profession. To get that reality back, one has to do some different things. I try to do it by using the simulation process. My class "lives" for a whole semester in Los Angeles County. We have our own city planning and zoning commission, among other things, and we fight a lot of battles. Many of my students know as much about the county's problems as the professionals. But we need all the outside help we can get. People from all over the country come to help us and send working material. I have a special library reserve in which I have accumulated over 1400 game-related documents. That reserve now is so complex that I am trying to computerize it in order to be able to more efficiently extract information. Even so, we do not have enough material. We have only one copy of the Los Angeles County North Section general plan, but we could effectively use six of them. Most of all, however, we need personal inputs. We need people who will work directly with us and will help us to understand what one can or cannot do in a given situation, and why, so that we might learn the "realities." It is a two-way cooperative street, and many academic people will be most receptive to your willingness to participate. There is a desperate need for more understanding, but you will have to work with us on it.

QUESTION: If the code process is disruptive to the good builder and comes about because of the actions of the bad builder, would it be more cost-effective to license the builder and treat him as a professional?

MR. DINEZIO: The legislation in Massachusetts was enacted in 1972, and our code went into effect in 1975. There were provisions for licensing construction supervisors, those involved in structural changes made in the building. We never implemented that program, however, simply because there was never any money for it. When you talk to home builders and the Associated General Contractors about licensing, you find that half are for it and half are against it. I believe that licensing is one of the best available methods for providing consumer protection because the building official can take an action. If I have a license and then do some sloppy work, the official has some recourse at license time and that means I will not be building in his town. There now is no way any builder can be stopped in Massachusetts even if he is one of the bad builders. Licensing is not a panacea, but it is the next best thing.

MR. LEDERER: We are trying another tool now. It has not been tested legally; however, we are proceeding with advice of counsel. With this approach, the building official can move if he is willing. If a builder is continually performing badly, we have established a hearing process through which the director is given the authority to stop the builder's permits. The rules of building are not changed, but attention is focused on the bad builder. He is stopped up front and, if he does not like it, he has the right of legal redress. We should not change the minimum standards of good construction that standards and codes set just to get at a bad contractor. There are other ways of doing that.

QUESTION: As a national contractor, I believe that licensing goes from the ridiculous to the impossible. In some cities and states, all you need to do is hire someone with a license to be automatically approved as a contractor. In others, the tests are impossible. The situation seems very political, and it allows some cities and states to have contractors who lack good judgment. Can the federal government do anything to make the licensing laws somewhat more uniform?

DR. BROWN: I am a member of a profession that is licensed, and I agree that it is a rather laughable process, at least when you try to go from one state to another. The medical profession is licensed as well, but I doubt that many of you would be willing to depend solely on that licensing process and to give up your right to institute an action for malpractice. I am sure the professions would be very happy to put in even more stringent licensing requirements if you would permit them to escape malpractice actions. I do not think you really trust doctors enough to do that, and I am not sure it would work for anybody else either.

QUESTION: Some states have created a one-stop permitting station. This is an effort of local governing bodies to speed up the permit process. What do you see as the future of the one-stop permitting process on a statewide basis?

MR. LEDERER: Our state provides a plan review service at a charge for those who need it. Once a plan leaves the local level, I question how effective one-stop service can be since the local official has lost track of it. The key to a good one-stop service is control and responsibility. I doubt whether you can control a process by sending plans by mail.

MR. DINEZIO: Massachusetts is going to try to establish a one-permit process, but I do not think it will work. There is a home rule problem that must be considered. No local entity will let the state take away its authority to that extent. I do, however, think one-stop statewide permitting will work if it is enforced on the local level. The state needs to establish its regulations and provide technical assistance to cities and towns to make sure the regulations are being carried out. We now do this with the appeals process and with educational seminars.

MR. HAMILTON: It would be terrible if every builder had to go to Sacramento from Los Angeles to get a building permit. There is, however, another alternative. A number of state agencies have some regulations governing building or development. I suggest that the state level needs to identify the statewide issues and concerns and then decide at what level those concerns should be exercised. Some of them probably could be exercised at the local level, the county level, or the regional level. Maybe none of them actually would have to be controlled by the state, but they would be dealt with by state legislation. The same should happen at the regional and local levels, and responsibility should be delegated to the level that will make it easiest for the builder or developer to get his permit and that will facilitate review, enforcement, and inspection. Our biggest problem is that we do not decide what is really needed and what we are trying to do. At the state level, you certainly can have one coordinating point and then draw upon the various necessary skills to save the builder or developer a tremendous amount of time.

DR. SIMONS: One-stop permitting has particular relevance to industrial projects whose construction falls under the purview of federal legislation. In site selection, for example, the ultimate choice may be dictated in part by the fact that the state in which you have chosen to construct a new plant already has taken all the necessary steps to assume responsibility for federally mandated programs. In this case, all the permits required under environmental laws would be obtained from one overall state regulatory body and not from both the EPA and the state. I also would like to endorse what Mr. Lederer said about the key importance of the pre-application conference. General Electric emphasizes that the regulatory agency should be contacted as soon as relatively firm plans have been made. We have found that we can avoid headaches later if we lay it all out for the agency early and determine if there is any fatal flaw from a regulatory point of view. Those responsible for the GE plants with the greatest potential environmental problems have learned that is the most prudent way to proceed in order to get approval of a plan in an uncertain regulatory field. There are risks, however, even when you go this route. For example, Consolidated Edison is starting work on a $150 million project to convert three oil-burning plants in New York City into coal-burning plants. The company has told the regulatory agencies that they need approval because oil will be in short supply. The regulatory agencies, however, feel they do not have enough data to give approval. Thus, Consolidated Edison is putting $150 million on the line because it feels that the risk of another Arab oil embargo or a curtailment of oil is greater than the risk of losing $150 million if the permits ultimately are denied.

MR. DINEZIO: We have talked about one-stop permitting, but I do not want to leave you with the impression that state governments do not have some problems. On the other hand, one-stop permitting makes sense if you think in terms of a construction project that is coordinated by a general contractor. In my opinion, the building regulatory system should function in much the same way. The various inspections should be consolidated in one state agency. That also should happen at the local level. In many cases, however, the building department and the building commissioner are responsible for one thing, but not for electrical and plumbing work and so forth. In such a situation, there is no coordination on the issuing of a permit. My point essentially is that the more a state gets involved in coordinating activities, the more readily consolidation will occur and that will expedite the permitting process.

Session IV

HUMAN MOTIVATION AND INCENTIVES FOR IMPROVING PRODUCTIVITY

PANEL OVERVIEW

R. THAYNE ROBSON (Moderator)
Director, Bureau of Economic and Business Research, and
Professor of Management and Economics, College of Business
University of Utah, Salt Lake City

The topic of this session, human motivation and incentives for improving productivity, was explored as were the other session topics, in workshops held early in 1979. It is my assignment to focus your attention on some of the issues that were raised during those workshops.

I assume that no one will take issue with the suggestion that one important key to increased productivity in the construction industry is to be found in the development, utilization, and performance of the men and women who work in this industry. Indeed, another way to view productivity is to examine the construction process from initial project conception through delivery of a completed project. Any blame for poor productivity in this and other industries is usually attributed to human failure in management or in worker performance. Yet is it appropriate to place this problem of productivity in the American construction industry in perspective. Few, if any countries, including Germany and Japan, can out perform the American construction industry. American firms compete successfully throughout the world whenever they are free to do so.

I also want to add my voice to that of those who are skeptical about the precise validity of our productivity measurements over time, either in terms of outputs or inputs. The outputs of today's heavily regulated, environmentally sensitive industry are not the same as those of the industry 20 years ago and the inputs are not the same either. Measurements of such things as hours worked and product quality require careful study before too much is made of productivity data.

These points aside, those who have worked in or studied the construction industry generally believe that it has seldom been considered to be a model for other American industries to follow in the development and utilization of human talent even though it has made significant strides in new methods, new technology, and new materials. Thus, it is appropriate to address the problems of human motivation and incentives as they apply to increasing productivity in this important industry.

At the risk of being trite, I ask each of you to think of your own career and the careers of others, whom you know well, in design and engineering management or in the skilled crafts. Now, ask yourselves some very important questions:

1. Has your productivity or job performance ever suffered because your boss or supervisor did not know or care about your job performance and could not fairly appreciate or evaluate your work?

2. Have you ever had experience with a boss or supervisor who could not or would not communicate clearly the job to be done or the full range of factors that would influence performance on the job?

3. Have you ever been the victim of inadequate training in the managerial functions of your job (i.e., planning, organizing, staffing, supervising, and controlling the responsibilities assigned to you)? (My discussions with people in the construction industry very often reflect the proposition that learning too often occurs after the job is done.)

4. Have you ever experienced conflicts between contractors, unions, inspectors, or engineers concerning when and how a job should be performed?

5. Have you ever been the victim of improper planning or scheduling or poor tools and equipment?

6. Have you ever experienced a situation in which project management dealt inadequately with environmental factors (e.g., parking, eating facilities, places to change clothes, and orderliness of the work setting) that it could improve?

7. Have you ever experienced an unnecessary union-management conflict that resulted in some subtle and some not so subtle undermining of job performance by both project managers and workers?

8. Have you been the victim of a set of attitudes about the industry, the people who work in the industry, and the unions that attribute evil or unhealthy motives to those with whom you disagree?

This list of factors with which you may have some personal experience hopefully will help us to highlight the importance of the human interactions required if productivity is to be improved.

In the brief time allowed for my overview of important issues, I want to present some modest suggestions for improving productivity. I will comment on four possible areas for reform that I believe have tremendous potential for improving overall human performance in construction.

First, I will address the broad area of training and education, including retraining and re-education. An entire forum could be devoted to this subject, but a few observations will have to suffice. It is essential that training and education be viewed in a broad context, not simply as a skill development process for both managers and workers. Training and education opportunities provide knowledge, skill, and motivation.

If I were to single out the two greatest training needs faced by the industry, the first would be managerial training for all levels of supervision in construction companies. I continue to encounter a large number of contractors who are former craftsmen and who believe that they can manage relatively large organizations with an engineer, an accountant, and a lawyer so long as they maintain a mountain of suspicion about the usefulness of each. I realize that important steps are being taken to build new college and university curricula that should have payoffs in years to come. What is lacking is ongoing training programs to improve the managerial skills of project superintendents, engineers who become managers, and the thousands of small and medium size contractors (perhaps 75 percent of the industry) throughout the country who already are managing a $200 billion industry.

The second major need is to provide craftsmen with the opportunity to review and upgrade their skills and to be informed about factors affecting the industry (e.g., the outlook for employment, regulation, new technology, and new materials) so that uncertainty and threatening surprises arising from lack of information about the industry and its opportunities for craftsmen will be eliminated to the extent possible. If this were done, the resistance to training of more apprentices would be less intense.

The second area in need of reform is planning and scheduling of work. Successful scheduling presupposes careful and detailed planning that is based on thorough knowledge regarding the site and the availability of labor and materials and on recognition of the different problems that may be encountered by different subcontractors.

I recently have been reviewing projects of medium size and larger. The major delays that occurred on these projects resulted from conflicts that arose when subcontractors failed to meet delivery schedules and negotiated amended schedules, thereby throwing other subcontractors and their crews and the whole project off schedule. This is a severe blow to human motivation and stimulates an unhealthy increase in conflicts between contractors and unions. The only beneficiaries are the lawyers who threaten the lawsuits and the people who usually benefit from substantial cost increases. Labor turnover and absenteeism increase and the general level of hostility in and around the project is damaging to productivity.

The third area for reform involves better union-management cooperation. It is clear that we know some things we fail to implement in our management of the industry. Workers are loyal to the unions, to the crafts and to the industry. Most of the available research shows that the majority of workers take pride in and gain personal satisfaction and fulfillment from their work and desire to do a good job most of the time.

In this respect, their behavior is the same as that of managers, architects, engineers, and even college professors. However, the persistence of perceived threats to any of these loyalties and constant exposure to unnecessary conflict can erode performance and satisfaction. An environment of cooperation recognizing joint labor-management goals can do much to facilitate worker performance. The actions required to achieve cooperation can best be worked out at the local level. Although these can be supported at the national and regional levels, they cannot be forced.

Among the things that should be used more widely are pre-job conferences, better employee facilities, commitments to prompt and timely conflict resolution by management, project labor agreements, jointly sponsored and managed training programs modeled after the best of the joint apprenticeship programs operating around the country, better written and oral communications, and better worker and management performance reviews. Some of you have been involved in both project reviews and evaluations of worker and management performance. Much more could be made of this technique because the best time to resolve a problem is immediately after it has been encountered, not weeks or months later when it arises again.

We may all be surprised at what emerges from the formal cooperation programs now under way in at least half a dozen cities in this country. If a cooperative environment existed and if conflict management procedures were accepted and trusted, some of the long-standing conflicts about restrictive practices could be resolved successfully. Surely, a joint management-union approach to unnecessary or inappropriate government regulations would meet with almost certain success.

My final suggestions for reform relate to basic human attitudes and human values. The T. S. Eliot quote used by Dr. Segal is a proper introduction to what I have to say: "We cannot create systems so perfect that no one will need to be good." There has been at this Forum and at other conferences on the construction industry a great deal of criticism of self and others, of unions and management, of architects and engineers, of owners and regulators. Interesting statements are made about workers, managers, and owners and about their wages, profits, and behavior that imply two things: that someone really understands their real motives and that their real motives leave a great deal to be desired. We too often both praise and condemn expressions of self-interest and of self-aggrandizement. The construction industry, while facing boom and bust cycles over which it has only limited control, has a bright and challenging future. Any of the conceivable solutions to our energy problems will entail sizeable amounts of construction. Likewise, expanding our military defense system would increase construction activity. The capital investment needed to increase productivity in other sectors of the American economy will require new and modern industrial facilities, and the demand for housing will remain strong. As you all know, however, the demand for construction is a function of the cost and availability of money, and the cost and availability of money is heavily influenced by the oil exporting countries.

It is a paradox that the new and emerging policies and practices in human resource management are developing in many companies that are large customers of the construction industry but that these emerging practices have not been adopted within the industry. Improved productivity from better human efficiency may require significant attitude adjustment, and it may be that the key to better training, planning, scheduling, and cooperation is to be found in the basic attitudes that we hold about ourselves and others in the industry.

PRESENTATIONS

NEIL B. McARTHUR
Vice President
The Austin Company, Washington, D.C.

Human motivation and production in the construction industry have always been problems, and we work at it constantly. Austin is a big national company. At any given time it is involved in about 100 projects in the United States and another 15 or 20 in Canada; therefore, we think we know how to motivate people and how to get production.

The key to production is scheduling and scheduling is a huge problem in our industry. To try to alleviate this problem, we run an annual week-long training session for our schedulers. Scheduling is vital to production and it also is vital to selling a job. How do you convince an owner that you can complete a job when he wants it? We must develop a schedule that the owner can understand. In addition, we must set up another schedule that the project superintendent can understand and work with and still another that the engineers can understand.

Manpower is another of our serious concerns. Our biggest problem in this regard involves having our materials on the job when we have the manpower available to use them. Many factors are involved, and anything can happen and constantly does.

Tools and equipment also are important. Poor quality tools or the lack of a sufficient number of tools has a tremendous effect on construction worker productivity. On the other hand, however, we have a serious problem with respect to the pilfering of small tools. The cost of such pilferage is great, and we have tried many methods to control it but have not been very successful. Heavy equipment is only a problem in terms of scheduling. You must think ahead if you are going to need a 250-ton crane so that you can plan to have it available when needed.

The Austin Company is constantly training in two areas. As an engineering and building organization, Austin hires all types of engineers. Often those coming to us straight from the universities must be totally retrained to our particular method. They have all of the theory, but they have no ideawhat a construction job is like. In our training program, we try to expose these engineers to our jobs. After a couple of years of training we send them out as field engineers, which often means that they have to move to the construction site where the engineering jobs are. Of course, many wives do not like that so we often lose good engineers who we have spent considerable time training. Thus, mobility is a serious problem, and it can be difficult to get professional engineers for the average two- or three-year construction job. The situation is different for huge petrochemical and power projects that can last up to 15 years. We can set up training programs, and people will move for such long-term work.

One way of increasing production on big jobs is to industrialize much of the work. If we establish an industrial-type operation using production line techniques, workers require less skill and are easily attracted because they will be employed almost permanently. We are, in fact, using more and more pre-engineered, pre-built materials that are delivered to a job as complete units. By industrializing certain segments of big power projects, we can increase production and better utilize our manpower.

We also are doing more to train our field personnel. Presently, of about 150 superintendents, 120 to 130 have come up through the ranks of the various trades. Coming up through the apprenticeship ranks is managerial training and always will be. These are bright young men and women. We give them a lot of technical training and move them into foremen and then general foremen positions. If we were to train professionals for these field management positions, it would take 10 to 12 years because they just do not have the knowledge of the field. They generally do not know how to motivate or how to work with people. They do not understand labor problems or know how to deal with the union business agents or the union workers on the job. This can cause constant labor problems. When dealing with the unions, I always emphasize that neither side has a monopoly on the "bad" guys.

Another serious manpower problem is due to turnover and absenteeism, both of which are very expensive. The current breed of construction worker moves on for all kinds of reasons. He will quit if he does not like the foreman or the job or if he finds a job closer to home. He will not show up if the trout are running or the hunting season opens. This presents special difficulties since we usually work people in pairs. If one does not show up, you may or may not get another man. One of the big companies conducted a survey on a very large job and determined that absenteeism cost $0.50 per hour in terms of production losses.

Some of the unions are trying to work out a system to penalize people who are absent without a good reason and this might help. Labor and management jointly have set up the rules, and the unions enforce them through their control of the hiring halls.

Indeed, without the hiring hall, the big national contractors would be at a tremendous disadvantage. The greatest asset we have in being union contractors is that we have a source of competent manpower available. We do not have to spend months finding out if there is sufficient competent manpower available and what they can produce as we do overseas. For example, in Puerto Rico, wages are probably a third of what they are in the United States, but our cost per unit of construction there is the highest we have because the available labor lacks training and motivation.

Unfortunately, we contractors tend to treat a construction worker like we treat a compressor. When we do not need it, we get it off of the job. A little more humanism would be nice, but it is difficult given the way things work now.

Earlier speakers spoke about pre-job conferences and getting labor involved in planning a job. That might work on in a huge, long-term powerhouse or petrochemical project, but it would not be possible to involve labor at that stage on a normal construction job. First, the owner would not appreciate it, and he is the one who pays the bills. In addition, as a design-construct firm, we are still designing as we are getting ready to build. We know in general what we will build but not who the subcontractors will be or what kind of outside skin will be used. If I met with the business agents for the bricklayers, the carpenters, the pipefitters and the electricians, someone would ask about the outside skin. Since I would not yet know, the bricklayer would argue that we should use brick and the carpenter would argue for something else. After all, these individuals represent their trades, not the industry, and jurisdictional problems would arise even before we started. Once most of the preliminary decisions are made and the job is ready for construction, we do call a pre-job conference. At this point, we try to project our manpower needs for each trade and tell labor the approximate dates when we will need each of the trades. We also bring in the local business agents to discuss problems and benefit from their inventive ideas.

In closing, let me say again that we have to work at productivity. We have some very real managerial problems, and the industry is changing very quickly. We must consider module-type construction and recognize that more things are going to be made in factories. This will result in the need for many bargaining agreements, and we are going to have to work things out with labor. Those of us who are union contractors are faced with new, stiff competition from a growing non-union sector in the construction industry. There are lots of benefits I would like to see union workers get but somebody must pay for them. If the client is not willing to do so, where will the money come from? If we put it in our costs, a client will reject us in favor of a non-union company. The union can negotiate a wonderful contract for its people, but if we cannot get work under it, it will not benefit anyone.

The secret of productivity is a smooth-running job. Just one work stoppage will slow down a job, and it may take months to get the momentum back. A serious accident also can slow down a job considerably but will not have the same effect on momentum that a work stoppage will. Thus, one must have a good relationship with labor. One must establish a system for quickly resolving jurisdictional disputes, contract disputes, and any other problems that occur. We must remember that we are dealing with human beings and that their individual problems will carry over to the job. I rarely have trouble with jurisdictional problems on huge jobs that involve hundreds of hours. It is the half-a-day problem that causes me trouble and causes so much argument that I lose far more than a half day's work debating the principle of the thing.

F. S. KELLSTROM
Chairman of the Board
Fischbach and Moore, Inc., Los Angeles, California

I have been asked to speak from the subcontractor's point of view and to suggest positive measures for improving motivation and providing incentives among the craftsmen involved in the various subtrades.

We are all very well aware that slipping productivity is contributing in a major way to our inflation problem. Now, what can we do about it? At first glance we might be inclined to talk about or think of money as the greatest incentive to increased production. I do not, however, agree with this premise. Construction is not done on a piece-work basis. The construction worker is not paid by the piece he produces; he is paid by the hour. In many areas and in many local unions, a labor agreement clause makes it illegal to pay over the prevailing wage scale. I agree with this clause because it prevents contractors from pirating men from their competitors by offering to pay more than the prevailing wage. The same rule applies to the use of overtime to attract men. Sometimes it is absolutely necessary to use overtime to complete work on a project (especially critical military projects) on a given date, but in many instances, a contractor will put a fee job on six 10-hour days and will attract all of the good men in the area to the detriment of local contractors who probably are doing firm price, hard dollar work. I think this is very unfair, and I do not believe we should use such a practice.

If we cannot use money or overtime to provide incentive and motivation, what can we do? Our panel chairman mentioned pre-job coordination meetings, and I think these meetings are good. Most good building contractors have a pre-job coordination meeting, but some do not, and it is up to our supervision to insist that such a meeting be held. There are almost always potential jurisdictional problems among the various crafts that have to be settled before the job starts if work stoppages and slow downs are to avoided. (Incidentally, in spite of what Mr. McArthur said, a slow down is worse than a work stoppage because you have to keep paying during a slow down and that gets expensive.) If a stalemate is reached in negotiating some of these jurisdictional problems, the matter can be referred back to the craft union heads and a work assignment can be made. This practice saves countless hours of bickering and floundering on the job site.

It also is important to have the best tools and equipment. We have found that it is all-important to begin with and to maintain throughout the project either new or completely rehabilitated tools and equipment. Men simply produce more with good tools; they do not like to work with faulty equipment. Given current labor costs, it only makes good economic sense to furnish them with up-to-date and good working tools and equipment and to keep them in that condition. Many transmission line contractors maintain that to put marginal equipment on a transmission line job is to commit financial suicide.

Many of these jobs are in remote and practically inaccessible areas; therefore, countless man-hours can be wasted waiting for parts and repairs.

Many hours in the field can be saved if good field drawings are provided. An experienced engineer will study the various drawings pertaining to the specialty trades and will lay out his work in such a way that no craftsman has to waste time wondering where to install the electrical or plumbing equipment and piping in relation to the air conditioning ducts. When these drawings are available, the men know where the equipment goes and immediately go about installing it accordingly. This eliminates later modifications, which we all know are very costly.

Proper placing of men also is important. All journeymen are experienced in the various phases of electrical work, but usually one is better doing one type of work than another. For instance, it would be ridiculous to put a good control man on a pipe bending bench. Many journeymen like pipe work, and they are very adept at it. You really get your money out of them when you put them on a pipe bench, and it would be very silly to put them on a hard, tough, quick control job. Although they would get through it, it would cost you more money.

You also must weed out the nonproducers. There are some men within the craft unions who not only do not want to produce themselves, but also will do all in their power to keep other men from working effectively. Their primary purpose in disrupting the job is to slow it up and force it into overtime. We call these nonproducers "renegades" or "wobblers," and they can disrupt a job in many ways. They may start jurisdictional disputes, heckle the workmen who are trying to produce a day's work, and hold meetings on the job on the contractor's time. Fortunately for the contractors, the trade unions in general and the electrical union in particular have always been helpful in getting these people off the job. This should be done promptly because once such a type gets a foothold, he can ruin a project and any hope of labor productivity on the job.

I also want to talk about giving men the recognition they deserve and, as Mr. McArthur put it, not treating them like a compressor that you send back after you use it one day. These are highly skilled and experienced craftsmen. Let them know when they do well. Give them the recognition that they are due. Most of the managers and top people at Fischbach and Moore have come from the field forces, and I believe this is a very real incentive for the men. They know they are appreciated and have a chance for advancement.

The success or failure of a job depends most heavily on the supervision. If you select your supervisors properly, they will be able, very intelligent men who know their business and enjoy the respect of the men with whom they are working. They are leaders, not drivers, and as leaders they will get more work out of the men. If we are more selective in our supervision, we probably will do well as contractors.

ROBERT F. SCHMITT
President
Bob Schmitt Homes, Inc., Strongsville, Ohio

Our presence at this meeting indicates that we believe the construction industry has a productivity problem or, to put it another way, that construction costs too much. This can easily lead to the simplistic conclusion that if the construction worker, the manager, and the supervisor could work harder and smarter, productivity would be improved and the cost of buildings would decrease.

I do not believe this is really our problem. The designers of this conference chose to deal with the problem by addressing five aspects of productivity: the management process, financial planning, government and public action, human motivation, and innovation. Four of these five are management functions; the other is a public or governmental one. I am supposed to speak about human motivation and incentives. I am glad I can say that I believe human motivation and incentive to be alive, well, and indestructible. This is particularly true in the construction industry. Construction work probably offers more discomfort and less security than almost anything else; yet, the industry and its workforce are extremely resilient.

In my own business, I repeatedly stress that a basic human desire is to be looked upon with respect by one's fellow man, to have economic security, and to enjoy one's life, including one's work. I also stress that the effective supervisor must have the willing support of a majority of those individuals working for him.

In short, productivity is a management function, and without individual incentive, all of the other facets of productivity cannot and will not occur. One of the chief factors contributing to the productivity decline in construction, if that really is the case, is the deterioration of the management function. This deterioration has occurred because government regulation has deprived management of the latitude it needs to seek new and improved solutions.

When asked what I do, I answer that I am a practicing socialist. I build homes in a way that the government prescribes in quite great detail, on land that I can use and develop only in the ways the government permits, and with money that the government turns on and off at will for reasons other than the industry's existence. I am allowed to earn a fee for doing this, but I am not guaranteed one.

I have a little trouble maintaining my motivation in view of the current concern over Chrysler Corporation, which directly employs 150,000 people and has annual sales of about $16 billion. The government is getting ready to underwrite Chrysler for $1.5 billion while also destroying 1 million jobs and $60 billion in new home sales by permitting the Federal Reserve to withdraw credit by manipulating the interest rates on home mortgages. This is based on the 50 percent reduction in new home starts that already has occurred. This same credit crunch applies to nonresidential starts as well and probably has an overall economic impact on companies and jobs approaching that of the new home situation.

What kind of logic is that? What does it do to my motivation and that of the 100,000 others like me? What does it do to the motivation of the people who work for my little company and the 2 million other construction workers? What happens to our productivity growth when the government chooses to regulate its economy by toying with the jobs and the lives of the 2 million people who perform direct labor on 2 million homes per year?

Then, of course, there are building codes, zoning codes, and the political aspects of our business. What happens to motivation when you are not permitted to try anything new? I have meekly submitted a proposal for a passive solar house to my local community. It violates every provision of the building code, but it will work. I at least would like to have the chance to fail, but my proposal is simply laughed at. We have been sterilized by our own government.

I like to think of myself as a pragmatic, cynical curmudgeon, but I am not or I would not be here. If I did not believe in the construction industry, I would not devote so much of my time to work for the board of the National Institute of Building Sciences and for many other organizations. Given our media mentality, it is difficult to see the good of us in the construction industry, but all we need to do is look around. Look at the United States and then look at the rest of the world. Look at the standard of living in housing that we have in this country as compared to any place else.

We have reached the point at which the varying factions within the construction industry have more interests in common than they have opposing interests and they must join together to demand a reform of regulatory climate. The National Institute of Building Sciences (NIBS) was founded by the Congress on that premise. Whether or not NIBS can fulfill its purpose depends upon you in the construction industry. If you want NIBS to succeed, it will. If you do not, it will not.

We have problems, but we can solve them. We have great capacity to solve them. I may sound cynical much of the time, but I would like to emphasize the greatness of our ability to solve our problems.

ROBERT A. GEORGINE
President, Building and Construction Trades Department
AFL/CIO, Washington, D.C.

The construction industry has many detractors and critics who, when confronted with the truly human approach to motivation and incentives, very quickly point at a myriad of indices, some of which are used by Quinn Mills in his book, The Construction Industry (1979). These indices supposedly measure developments in the various industries, and from these, economists and statisticians feel confident in making projections concerning how a particular industry will fare, given certain economic and political factors.

The current issue of the National Planning Association's magazine, Looking Ahead, offers an article by Jerome A. Mark, Assistant Commissioner for Productivity and Technology at the Bureau of Labor Statistics, U.S. Department of Labor. He states:

> What has been happening to productivity? In 1978, productivity as measured by output per hour of all persons in private business economy, the largest sector for which we have aggregate measures, rose 1.1 percent, extending a decelerating trend which has been taking place since the mid-1960s. And from 1947 to 1965, it has risen about one-half that rate, to 1.8 percent a year. The decline since the mid-1960s, however, must really be broken into two periods; the period from 1965 to 1973 and the period from 1973 to 1978. And the factors affecting productivity in each period are quite different.
>
> Moreover, the economic sectors which contributed to the deceleration of the first period differed from those in the most recent period. From 1965 to 1973, the trend rate dropped to 2.2 percent per year and from 1973 through 1978, the rate dropped to 1.1 percent per year.

I will not challenge those confusing figures, but I will present the observations Dr. Mills makes in his book:

> Labor productivity is said to be low and the industry has generally held to be costly and technologically stagnant.... Bricklaying, for example, is said not to have changed in thousands of years. Perhaps, in the literal placing of a brick on top of another brick, it hasn't. But masonry technology has changed a great deal. Each of these advances has made masonry a more modern and efficient industry and has enabled it to compete on a cost basis with such indisputably modern technologies as steel and glass construction. Government statistics suggest that since construction does not use assembly lines, its technology is stagnant. Are these statistics accurate or are they misleading? Measuring productivity requires the raising of at least three basic questions. One is, are the statistics accurate; two, do the statistics

measure what they claim to measure and, three, what interpretation is appropriately given to the behavior of those statistics?

There was a time when I would have wondered who cares. Well, for the past 15 years I have cared because, like it or not, building and construction trades people have been and are constantly being victimized by biased labor productivity measures.

How can I and the presidents of the building construction trades affiliates translate this kind of information to the person on the job who does not have a clue as to what is being measured or why it is being measured? Just what are we expected to do? We can challenge theories. We can try to impose reason where there is none. We also can raise some questions. Is there a direct relationship between the declining productivity rate and the increased activity of the business round tables and the users and the construction process? Is the further decline in productivity related in any way to the increased growth of the merit shop movement in this country? Is there any relationship between the extending decelerating trend and the increased activities of the new right and its anti-worker kinds of policies. I could go on and anyone with an opposing view could do likewise, but neither sufficient time nor resources are available here to properly debate these issues.

I would like to return to a very simplistic approach to human motivation and incentive for improving productivity. I suggest that wherever management has not abdicated its responsibility to some well-meaning, insensitive team, productivity has increased and real earnings--both, earnings for the worker and profits for the contractor--have increased. This cannot be accomplished, however, without giving the worker some very simple incentives. Some such incentives are proper job planning and full and complete pre-job conferences at which all of the planning is explained so that the worker representatives can ensure that the job site will be managed properly in terms of labor relations, employee relations, housekeeping, scheduling, supplies, and, most of all, communication with the people who are actually expected to produce the project. For example, why would designers and engineers not consider putting in the permanent roads to the job and parking lots for the cars when anticipated time of construction is four or five years? Coupled with this are the monetary concerns of construction people, and money has to remain at the top of the incentive ladder.

Since construction personnel, in general, are exposed to more different kinds of hazards and situations that cause them to lose time, the direct relationships between the firm's management and direct line supervisors and the employee also are very important as an incentive. Surprisingly, however, only the most sophisticated companies continue to have that kind of communication. One often can find the greatest degree of sophistication in the smallest companies in the country. A small contractor can be very sophisticated in terms of the relationship between contractor, job site, and employees.

Anyone attempting to motivate individuals using these incentives must approach the subject honestly. He must know the group with which he is working as well as the ultimate objectives of the incentives. If a firm determines to offer, for example, free parking as an incentive but immediately takes away the coffee break, it certainly is not providing an incentive. A firm that is determined to take something away for every incentive provided will not increase motivation; instead, it will nullify whatever progress has been made toward motivating the work force.

The building tradesman probably is the most maligned of all construction industry personnel. Nevertheless, U.S. construction workers are the most productive in the world, and U.S. construction companies are the most productive in the world. There is no question in my mind that good productivity will continue if we can eliminate some of the problems discussed during this Forum. Productivity or manual output is a very human thing, and a great source of knowledge for planning, for accomplishing the end result, is the worker who will be doing the job. Some of the greatest innovative ideas have come from bricklayers, pipefitters, and electricians. They see what they are doing as they are doing it and they develop ways to improve it. These workers also see terrible mistakes being made by supervision, and if they are taken into your confidence, they sometimes can be of great assistance. It is important that you not lose this asset, this contribution. However, management usually thinks it has all the answers and insists on telling the worker how to do his job most efficiently.

The contemporary construction process does have a built-in disincentive. Technology has advanced rapidly and construction now involves more pre-fabricating, pre-assembling, and modular work than it has in the past. Therefore, many of the skills needed by the individual worker today are not as great as the skills that were needed on the job site years ago. Much of the comtemporary worker's time on the job now is spent in connecting, assembling, and doing things that do not give him much emotional pride.

That is not to say that there is nothing workers can do to increase productivity, but, by and large, the things that will increase productivity on a job site now are good management and good planning--making sure that there is enough work for the manpower that is there to do it and making sure that the job is laid out in such a way that men are not standing around waiting for material. Most men come to a job every morning to perform eight hours of work and go home. They do not come to the job to sit around or to do work improperly. They come there to do what they are trained to do as well as they can do it, to get it done, and to go home at the end of the week with a sizeable paycheck. Most jurisdictional disputes and job stoppages can be avoided if job assignments are made properly and if the people who made them were knowledgeable. When a dispute does arise, there should be established procedures for resolving it, and they should be initiated as rapidly as possible.

CHARLES B. KNAPP
Deputy Assistant Secretary of Labor for Employment and Training
U.S. Department of Labor, Washington, D.C.

I appreciate the opportunity to be here and have gotten quite a few insights into the human factors that affect productivity in the construction industry by listening to the other panelists. I will discuss the problem from my perspective and in the context of my work for the Department of Labor. It is always important to remember that we approach questions from different perspectives and that our perspective is all important in terms of the judgments we make.

I do not think it is possible to look only at the positive factors affecting productivity in the construction industry. We must examine the issue in a more general context. For example, many of the Department of Labor's activities could be viewed, very strictly and narrowly speaking, as affecting productivity negatively, at least as productivity has been traditionally measured. This does not necessarily mean, and I would argue strenuously that it does not mean, that these are "bad" things.

Probably the best analogy to use here is the difference between what we might call tension and what we might call slack. The different perspectives that we bring to bear on various questions force us to argue about issues upon which reasonable people can disagree and upon which value judgments are the deciding factors. To the extent that we are involved in questions like this, we ought to debate them. We have a lot of different processes from the very micro level to the very macro level for trying to see all questions, but we must remember that these processes, in fact, reflect our different perspectives.

By the same token, I believe that in many systems, including the construction industry, things happen that do not necessarily have to happen--that is, there are things that can be changed to make everybody better off and nobody worse off.

What are the things on the agenda of the Department of Labor, whose mission is to protect and promote the interests of American workers? It is involved in a wide variety of activities that affect the construction industry, but, in judging these activities, you should remember that although the government plays a large and important role, it is really only a bit player in the process. The solution to problems lies largely with the industry and the workers themselves, and although it might not be the popular thing to argue these days, I believe many of the problems tend to lie within the industry itself. Some of the problems and some of the solutions lie with the government, too, but we must not overestimate the impact of government.

The programs that I am involved in are the employment and training programs of the Department of Labor, and I am blessed to have responsibility for the Bureau of Apprenticeship and Training. The very heart of the apprenticeship system has tended to be in building and construction, and there are many interesting questions about apprenticeship right now. I do not have answers to all these questions, but I would like to raise a couple that are important to those involved in the apprenticeship system.

The success of the apprenticeship system has been to produce broadly trained, rather than narrowly trained, craftsmen. In my mind, this raises a set of questions about how we train people for construction or, in the broader context, about how we train the American worker in general. I believe this goes to the nub of the human factors issue.

How does one impart skills? The apprenticeship system has been marvelously successful in doing just that. It employs a particular model. It takes time and it spends intensive effort on the individual in order to impart these skills. But how do we use that system? How can we model apprenticeship for those occupations that are not normally thought of as "apprenticeable"?

Another interesting question involves what is to be done with apprenticeship or construction labor in general during an economic downturn. It is clear that when high interest rates and economic downturns reduce the amount of activity in the construction industry, we tend to lose much of the human capital that is embodied in the workers.

Skilled workers tend to leave the industry to do other things during economic downturns, and many do not come back when the economy improves. That is obviously a problem in terms of productivity because it will be difficult to locate skilled workers who can perform adequately when jobs are available again. Bottlenecks occur and, perhaps, result in increased inflationary pressures, and we repeat the cycle again. We also have to worry about the type of training of skilled craftsmen and workers in the construction industry that goes on during an economic downturn.

Another thing that is of great concern is training outside of the apprenticeship system. We have, through the Comprehensive Employment and Training Act (CETA), spent large sums of federal money attempting to train people. One of the interesting questions is how we can best go about that training. There has been, I think, some tension with the construction industry and with organized labor in the construction industry over the types of training that have been done. Those are legitimate concerns that often have to be discussed at the local level. Both on the industry side and on the side of organized labor, legitimate questions concerning competition are raised. For example, if the Labor Department gets involved in weatherizing homes, which is something it is under extraordinary pressure to do, questions concerning competition with small construction firms that also tend to be engaged in weatherization arise as do questions about the training of workers in what traditionally have tended to be unorganized fields of endeavor. These are the kinds of questions that tend to be resolved at the local level, and we have approached the problem in various ways. We have tended to concentrate heavily on low-income housing, an area in which established firms normally do not become involved. We also have tended to try to use, for example, organized labor to help train the people who are working on the weatherization site.

The CETA employment training system really cannot be viewed by the building industry as a source of broad-gauged workers, workers that are skilled craftsmen. All it does is to interest people, to acquaint them with things. The sort of person who spends six months or a year on a Department of Labor weatherization project is not going to be the sort of worker who is needed to increase productivity in the construction industry. Thus, we have to look for other routes that will allow people to get the broad type of training that is necessary.

A number of other Department of Labor program areas bear heavily on this question. One is labor standards, a major responsibility of the Department of Labor. For example, the Employment Standards Administration is responsible for enforcing the Fair Labor Standards Act, the Davis-Bacon Act, and the Service Contracts Act while the Occupational Safety and Health Administration and the Office of Federal Contract Compliance Programs are responsible for enforcing other laws. These classic examples of the Department of Labor's role relate to the issue I raised earlier about there being one difference between tension and slack. These laws can, in a strict and narrow sense, negatively affect productivity as traditionally measured, but that reflects a mismeasurement of the output of the industry. If these acts create a safer and healthier work environment, that ought to be considered. By the same token, there sometimes has been what I would classify as slack involved in the enforcement of these laws (e.g., worrying about split toilet seats rather than cancer-causing agents). We need to remedy such situations so that we can concentrate on the tough problems.

The Department of Labor obviously is also very interested and actively engaged in certain aspects of labor-management relations. Most of our responsibilities in this area are, interestingly enough, rather informal. We have a number of programs under way, but we often get engaged in, almost on an appeals basis, fairly large labor-management disputes that spill over out of the Federal Mediation Service or the other mechanisms for dispute settlement. These disputes often end up on the Secretary of Labor's doorstep. We are interested in making sure that things are solved in a manner that is acceptable to both parties.

We also are involved in, on an experimental and demonstration basis, construction coordinating committees in three or four major cities across the country. We are quite taken with this experiment and have had remarkable cooperation from both labor and management. The concept is very simple. Labor, management, and government representatives from a construction labor market area attempt to jointly work out problems of interest to all three. Often the government is a large builder in the area and is causing problems rather than attempting to solve them. In such cases, the government can do a number of things. It can ensure that its contracts are spread over time to avoid a seasonality problem in terms of the work load or the types of ups and downs and bottlenecks I mentioned earlier. We are quite enthusiastic about these experiments, and we think that they hold a good deal of promise for us in the future.

The last thing I would like to mention is the need for better information on productivity. The Bureau of Labor Statistics is an agency of the Department of Labor and one of the major agencies responsible for gathering statistics on productivity. When I was working directly for Secretary Marshall, I spent some time worrying about the process for organizing concerns of productivity in the government. I came away with the feeling that our methods for measuring productivity tend to be more art than science.

Many interesting questions come to mind when you begin discussing productivity. For example, we normally talk about productivity relative to a fixed capital stock. One of the reasons that U.S. workers usually are deemed more productive than foreign workers is that they have more capital to work with. Should we correct for that? Should there be a mechanism to build that in? Another of our concerns is whether we really are measuring the same thing now that we were measuring 10 or 20 years ago. In addition, as I mentioned earlier, should not the existence of a safe and healthy work environment be factored in somehow? That is a social good, and if we ignore it in measuring productivity, we may come up with the wrong answer.

WILLIAM J. CURTIN
Senior Partner
Morgan, Lewis, and Bockius, Washington, D.C.

Perhaps you may wonder what a management-labor relations lawyer is doing here talking about human motivation and incentives for improving productivity. The reason for my participation may be that I specialize in labor relations law and spend a fair amount of my time dealing with construction industry labor relations; therefore, I will attempt to address a productivity area that is legal and is within the special competence of the labor lawyer.

When a member of the general public talks about productivity in the labor relations context, he thinks of stereotyped images of employers who speed up assembly lines or of unions or employees who resist technological change. Given my perspective, I will discuss a different aspect of the industry that impacts upon productivity in a most dramatic fashion. This is the phenomenon of the construction industry strike during the term of the collective bargaining agreement whether you have real or imagined disputes or whether the dispute involves the employees or workers or their union and the employer or contractor. To state it positively, I will discuss means for improving productivity through the development and implementation of effective disputes resolution procedures--agreements between labor and management that not only allow but actually require work to continue while disputes are resolved in a fair and impartial procedure.

Strikes occur much more frequently during the contract term in construction than in most other industries. The last comprehensive analysis of work stoppages in this industry, done by the Bureau of Labor Statistics, revealed that 57 percent of all construction stoppages between 1962 and 1973 occurred during the life of the agreement, not at what the public regards as the normal time (i.e., when a contract has expired and the parties are not able to reach agreement peacefully on the terms of a new agreement). This was the case for only 34 percent of stoppages in all other industries. As a result, during that 11-year period, the construction industry was responsible for about 25 percent of all the contract term stoppages in the country. These facts, of course, suggest that there was something wrong and that it needed attention. Although it certainly is true that there are many detailed ways in which to work on the productivity problem in the industry, it seems to me that nothing more dramatically or decisively lowers productivity than a work stoppage on the job.

Beginning in the late 1960s, I had the opportunity to become involved in this problem on behalf of clients and was exposed to the phenomenon that sometimes is called a stabilization agreement or a special project agreement. A project agreement is a single collective bargaining agreement. It is negotiated by a construction employer or contractor with representatives of all of the building trades whose workers will be needed on an upcoming job. This single agreement covers all of the trades involved on that project. Such special project agreements normally are negotiated for major jobs. The definition of "major" may be determined by the size of the job as measured in dollars or it may have to do with other special factors such as location, longevity, special technologies, or national interest. On the Trans-Alaska Pipeline, for example, the building trades negotiated a unique special project agreement to meet the unique conditions involved.

There have been many other such agreements, and they contain many innovative conditions that one normally would not think of as being related to productivity. A discussion of the evolution of such conditions in special project agreements would be very interesting, but I will limit my remarks to describing what I believe is an intelligent way to approach dispute resolution and enforcement of promise not to strike or lockout on the job.

It is a mutual promise. It is meant seriously and taken seriously, and the difficulty we have is in developing contract language that permits us to use our own procedures to resolve any stoppages or disputes which occur during the life of a project. The desirable dispute-resolution elements in special project agreements are:

1. A comprehensive no-strike and no-lockout provision,

2. An enforcement mechanism exclusively for the no-strike clause and contained within the same contract provision,

3. A general and all-inclusive grievance procedure, and

4. A separate jurisdictional-dispute-resolution procedure that also is subject to the no-strike prohibition.

Many traditional local collective bargaining agreements contain no-strike clauses, but these clauses are not the same as the comprehensive no-strike clause, which is a contract provision the parties mutually fashion because of their basic commitment to the prevention of interruptions on a job. Happily or unhappily, a simple statement that there will not be any strike does not do that job effectively under the laws in this country at the present time. Therefore, we have embellished the simple language to include the four essential elements just mentioned. The building trade unions have been willing to cooperate in this embellishment because of their essential commitment and, I think, their sense of responsibility for making the commitment work, for making it effective and enforceable.

The language negotiated over 10 years ago in the Walt Disney World construction project agreement serves as a good example of the first part of a dispute resolution procedure--a comprehensive no-strike, no-lockout provision:

> There shall be no strikes, work stoppages, picketings, or slowdowns by the unions or employees against any contractor covered by this agreement or a lockout of them while this agreement is in effect. The contractor may discharge any employee violating this provision. The contractor and the union shall take all necessary steps to obtain compliance with this article and neither shall be held liable for conduct for which it is not responsible.

This language contemplates that all work stoppages are prohibited. Subsequent project agreements have embellished further upon it. For example, the pertinent part of a nuclear power construction agreement reads:

> The union and its members, agents, representatives, and employees shall not incite, encourage, condone or participate in any strike, walkout, slowdown, picketing, sympathy strike, or other work stoppage of any nature whatsoever for any cause whatsoever.

The no-strike enforcement mechanism is the second clause that is critical in any effective dispute resolution procedure in a collective bargaining agreement. This no-strike enforcement mechanism is an example of the building trade unions and the construction industry at their most innovative. Procedures have been worked out in a number of special project agreements that really give the parties a mechanism to stop strikes, even mid-term contract strikes which often involve the employees striking or engaging in a work stoppage on their own without union sponsorship. Indeed, one of the reasons why building trade unions are willing to enter into effective dispute resolution provisions is that they do, in good faith, intend to negotiate and enforce the no-strike clause even if it must be enforced against their own members.

The procedure we have developed to deal with the problem of enforcing no-strike commitments is a device called expedited arbitration. Expedited arbitration is necessary for a couple of reasons. The law, although somewhat better in this sense than it was a decade ago, is a long way from being perfect in terms of permitting one of the parties, usually the contractor, to go into court and obtain a federal court injunction (i.e., a judge's order to striking parties to stop violating the no-strike clause). Expedited arbitration essentially involves our agreeing on some neutral person, who we think is knowledgeable, fair and available, to serve as arbitrator. In the event of a violation of the no-strike clause, the contractor has the option of either going to court or going to this special arbitrator, whose function is to conduct a hearing very promptly and to make the single determination of whether the no-strike clause is being breached. If the arbitrator decides the clause has been violated, he is contractually empowered to issue an order against the striking parties to return to work.

That order is enforceable in court in a way and to a degree that a direct appeal to the court would not yield. Now, that sounds just about like what lawyers are accused of doing to the system, but it is a fact and it is that simple. You can take the same no-strike clause, but if you go directly into court, you may or may not have a legal right to enforcement of it. However, if you go to an arbitrator first and he finds a violation of that clause, you can obtain enforcement of the arbitrator's decision in court rather easily. The result is the same. Obviously, the process is a little bit slower if you have to go to the neutral first but not meaningfully so if there is a question as to whether you can get an injunction by going directly to court.

In addition to enhancing the availability of a legal remedy, another reason and, at least in my judgment, the more important reason for the development of an expedited arbitration clause for resolving work stoppages is the fact that we are not talking about a procedure that is imposed by statute, but rather about something that the parties create, that they work out the details for in terms of their contract provisions, and it makes everybody understand how important, how central a piece, the no-strike clause is to the special project agreement. Working out the details about how you are going to get it enforced, if you sit there and bargain that through with each other, necessarily adds to one's understanding of how critical a factor the no-strike promise is to a special project agreement. And as a result, with that impressed fully upon the minds of the parties, these procedures have, for lack of a better term, served as a "deterrent" against breach-of-contract strikes.

One of the phenomena that we have noticed, not in all cases but in many, is that the incidence of breach-of-contract strikes on a job covered by a special project agreement with an expedited arbitration procedure has been remarkably low. That may be, and we hope it is, because the parties are more conscious of their commitment not to strike and of the availability of quick, swift, and certain remedy in the event they do.

The third essential clause involves the grievance procedure. It is, of course, extremely important to have an effective grievance arbitration procedure. To be effective, it must yield a resolution, hopefully a fair one, within a reasonable period of time. From the contractor's standpoint, having the capacity to obtain injunctions is not very useful if there are no mechanisms in the contract that he and the unions can use to resolve and diffuse underlying disputes.

Finally, the fourth clause to be mentioned is the one dealing with jurisdictional disputes. These disputes are the bane of the construction industry in terms of its reputation for stability and labor relations. They are still a real problem, but the situation has improved in the past 10 years primarily because of joint efforts by management and the building trade international unions to deal with something that was, at times, virtually a public scandal for the industry. There remains, however, a need for carefully written contractual provisions that provide for resolution of these disputes.

It is unfortunate that there are no statistics available on the number of contract term work stoppages occurring under special project agreements with comprehensive no-strike clauses so that we could verify the effectiveness of such agreements. However, there is no question that the number of mid-contract strikes in the construction industry as a whole has been decreasing during the 1970s. It is fair to note, I believe, that it was during this same period that the special project agreement containing sophisticated provisions came into pre-eminence in the industry. From 1971 to 1978, work stoppages in construction dropped by about 50 percent and work stoppages during the life of construction agreements dropped by 78 percent from 374 to 87. As you will recall, between 1962 and 1973, 57 percent of all construction stoppages occurred during the life of the agreement whereas in 1978 that was true of only 23 percent of the stoppages in construction. These statistics seem to indicate that the objective of the clauses I have described are being achieved for one reason or another.

Having covered the who, what, and why of the special project agreement no-strike provisions, let me devote a few moments to the question of how. How do you obtain comprehensive no-strike provisions? Why are unions willing to agree to such an innovative procedure or, for that matter, to the special project agreement itself? Of course, the contractor's reputation for good labor relations is a key. However, it is undisputed in the industry that the major share of the credit for special project agreements goes to the building trade unions. They have a long tradition of responsibility to their contractors in collective bargaining and they certainly have a commitment to the viability of the industry and their role in it.

In practical terms, the leaders of the building trade unions recognize the need for improved productivity in the industry, as evidenced by their participation in this conference. They are receptive to innovations, which they perceive as accomplishing increased productivity without unreasonable or unfair costs to their members.

The quid pro quo for the building trade unions' acceptance of the terms of a special project agreement is the contractor's guarantee that work on the entire contract will go to union members. The disposition of the unions to grant concessions is a function of their perception of a good faith commitment of management to encompass as much of the work as possible within the agreement they negotiate.

WALTER H. LEWIS, AIA
Professor, Department of Architecture
University of Illinois, Urbana

The subject of my presentation is public-private joint-venturing to improve productivity through better management of the community development process. Thus far, most of the panelists have concentrated on only one aspect of the community development delivery system--on-site management-labor relations. To put my remarks in the correct frame, you now must think not of the job site but of the earliest phases in the construction process: project initiation, project feasibility, project programming, and conceptual design. My argument is that more effective productivity will be the direct outcome when better incentives are provided to motivate joint public and private decision-makers during a project's initialization phase. I call this public-private joint-venturing. If you find this concept totally out of context with this discussion, remember that we just heard that improved on-site labor-management relations is the direct outcome of better labor-management contract negotiations conducted before construction ever begins. As we will see, improved management of the community development process requires new processes of public and private decision-making analogous to contract negotiations before design and/or construction decisions are finalized.

Before beginning, however, I would like to note that we just heard another statement that troubles me: that it takes 12 years of job-site training before graduate engineers and architects can function adequately as construction project managers. If true, the statement says more about the employers of new graduate engineers and architects than about the halls of academia that prepared them. Employers and academicians alike must decide what is best accomplished in school and what is best left to practical experience. Matters that require experience for their understanding just cannot be understood without experience! Employers have a responsibility to create the job environment that will provide, in a systematic way, the required on-the-job experience, which is impossible to simulate in an academic setting. This is absolutely necessary to assure that academic theory can be optimized in actual practice. Theory, by the way, helps us do two important things: (1) explain certain phenomena, and (2) predict the outcome of our decisions. Both are absolutely essential to a construction manager. The issue is not that we have too much theory (we can never have enough), but that we need better on-the-job professional development after graduation.

We also need to recognize that there is a difference between the skills required for engineering and architectural design and those required for engineering and architectural construction. In either case, the graduate engineer or architect ultimately will require management competencies. That is why the University of Illinois has a dual degree program in which one can pursue a master's degree in architecture and at the same time satisfy the requirements for a master's degree in business administration. I guess my only argument is that the practical experience phase should require only 3 to 5 years, not 12 years! Continuing professional education is germane to some of the conclusions I will draw later.

To get back to the point, the single issue that brings us all to this Forum is the inevitability of change. If we could predict the future to be the same as the past, this meeting never would have been called. Look at some of the current forces for change:

- Energy
- Ecology
- Equity
- Economics
- Employment
- Education
- (E)urgency

These key words emphasize that the decisions we make in the future can never be the same as the ones we have made in the past. Any session such as this involving our collective education requires change. In fact, that is what education is all about--changing our behavior. Today's program has been organized in response to changes in availability of resources, excessive costs, increasing regulations, and often-conflicting priorities.

Improving productivity through human motivation and incentives is not so easy. Although each of us recognizes the need for change, there are numerous reasons why we all resist it (and, particularly, why we resist the methods often used to effect change). For example, we resist change:

1. When the goals are fuzzy or when our purpose in relation to goals is not clear.

2. When our participation is ignored. (Never ask people to participate when their participation will not matter.)

3. When there is fear of failure because we fear what we do not understand. (Education can help overcome fear of failure.)

4. When the costs, not only economic but political and social, seem too high.

5. When we do not think our participation can make a difference.

6. When there is poor communication.

The last is the reason I want to emphasize throughout the remainder of my presentation. Improved communication is the major issue in achieving more effective productivity.

Others before me have described the fragmentation in our construction process. Problems exist and have developed because individuals and their institutions pursue their own interests. If we are to find new arrangements and new solutions in the construction process, we will have to find ways to bring self-interest to bear on the solution.

I want to argue for a stronger joint venture between public and private decision-makers to bring the interest of private development more in concert, if we can, with the public or community interest. To do so will require a new approach to public and private joint ventures through negotiation as illustrated in Figures 1, 2, and 3. Ignoring the complexity of defining a three-dimensional dynamic process on a two-dimensional drawing, let me explain. The bottom line in Figure 1, the public domain, represents the minimum standards below which we cannot go. The source of the standards is the police power to protect public health, safety, and welfare such as zoning ordinances, and building and housing codes. (That bottom line also reflects the source of the over-regulation so many have referred to in these discussions.) The top line in the diagram represents community goals. Community goals should be expressed through public policy in a comprehensive plan that has been developed by assessing the community's capacities. The way to achieve stated goals is by joint-venturing through negotiation. Public policy should involve spending our collective money on developing a much better data base about the capacities of the community to accommodate development. The public input should be to provide much better information about the community's natural capacity, manmade capacity, market capacity, and fiscal capacity. All of these factors add up to the community development capacity.

Earlier I was taken by Dr. Simons' point that, for the first time, General Electric has had to consider pollution laws in locating plants (i.e., the capacity of the air shed in the community where they have proposed to build a plant). This kind of information about the community's ecological capacity should be provided by the community rather than by the developer.

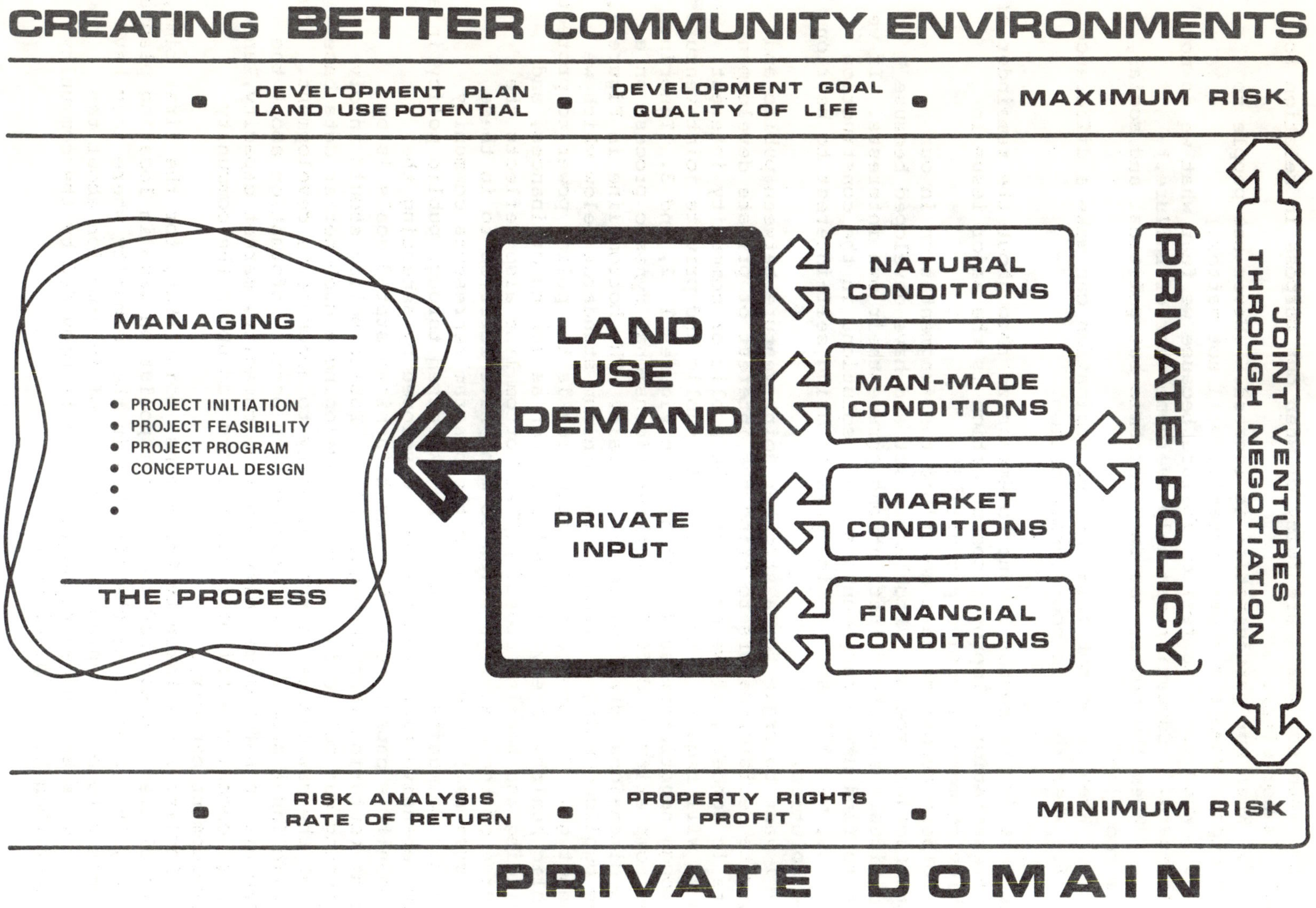

FIGURE 1 Creating better community environments--The Public Domain.

The bottom line in Figure 2, the private domain, has two parallels to that of the public domain; constitutional property rights that grow out of land ownership and the need to produce profit. Profit is on the bottom line because in our free enterprise system we are not in business if we do not make a profit.

As on the top line in public domain, we have a parallel on the top line in the private domain, the establishment of development goals and private policy. Private policy is determined by looking at the demand factors caused by natural conditions, manmade conditions, market conditions, and financial conditions. These four factors add up to community development demand. The process then in Figure 3 is to see whether private development demands can be balanced with community capacities. The problem is to establish equity in bringing private interests more in concert with community interests. If we truly want to accomplish community goals, we must find new ways to do so through negotiation.

A very important point can be made by looking at the parallel goals on the top line in Figure 3. The achievement of community goals requires maximum judgmental decision-making by elected officials, which is often very difficult to achieve; on the private side, we see that as we try to achieve maximum goals, the maximum risk is to the private developer.

These figures illustrate a more rational community development decision-making system. To be productive, delivery systems should get things done. They include the process of defining need, weighing alternatives, choosing priorities, and designing solutions. In reflecting on past development practices, community leaders are learning that many practices have been unproductive (i.e., incremental, costly, haphazard, and irrational). Development it seems just happens--the work of the "invisible hand." Clearly, community development is the work of developers. However, public developers (e.g., elected officials, appointed advisory boards, and hired professional staff) make innumerable decisions that have tremendous impacts on the outcome of the process.

The correct solution for the most productive delivery system is present in every community. There may be multiple answers, many are not always obvious, some are often elusive, and each will change over time. However, new systems for improved decision-making must be locally created and based on local issues and local opportunities. This means that each community may need a unique local system because each community has a unique market demography, legal framework, economic climate, leadership power structure, natural and manmade attributes, and past development patterns. The most productive local delivery systems are those that minimize the adversary role of the developer and maximize joint-venture opportunities by providing better assessment of development alternatives.

Even if the perfect delivery system could be devised, it would not function productively without enlightened management and leadership. A key idea is to initiate new local programs of leadership and management education. Leadership education can overcome the biggest

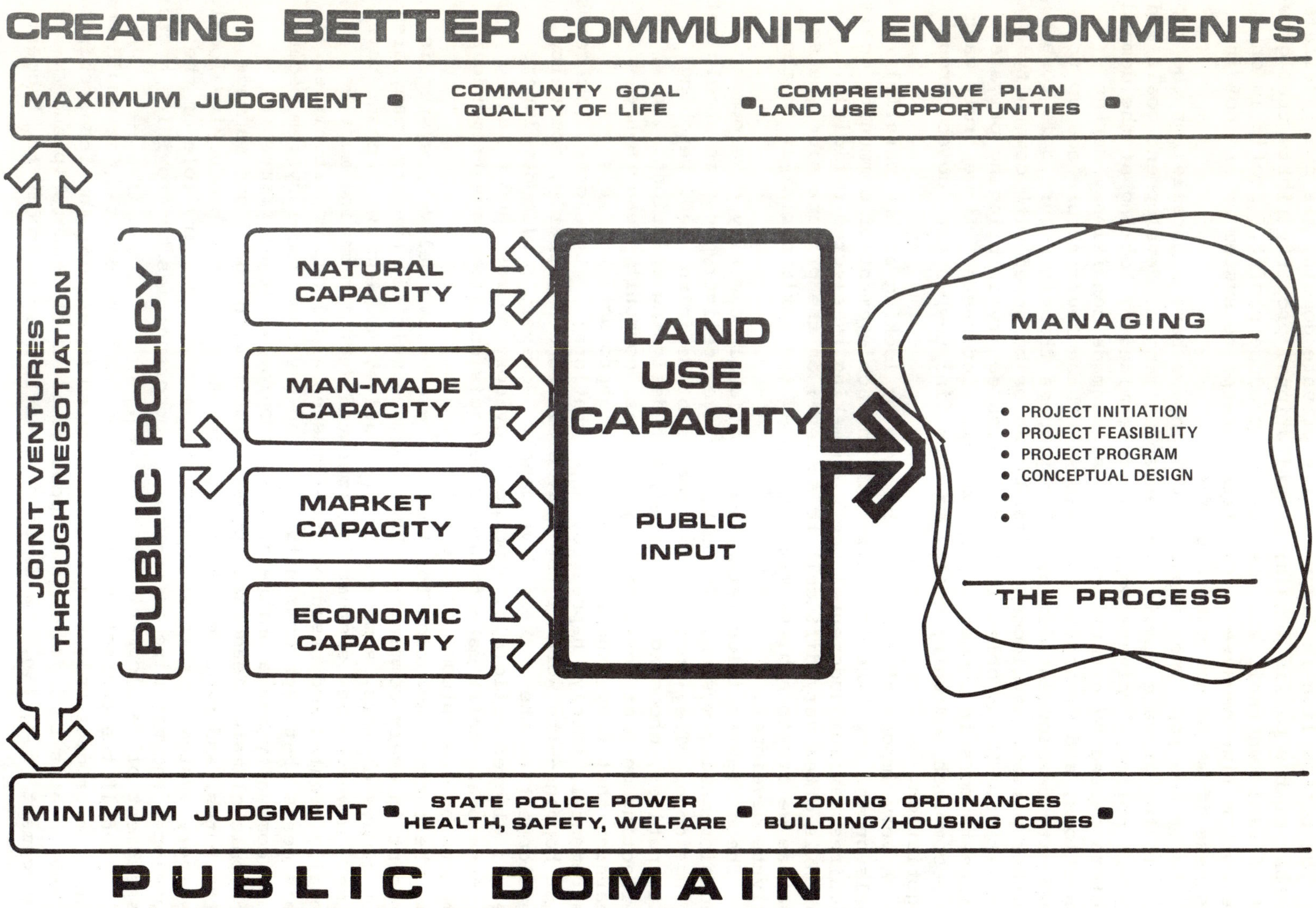

FIGURE 2 Creating better community environments--The Private Domain.

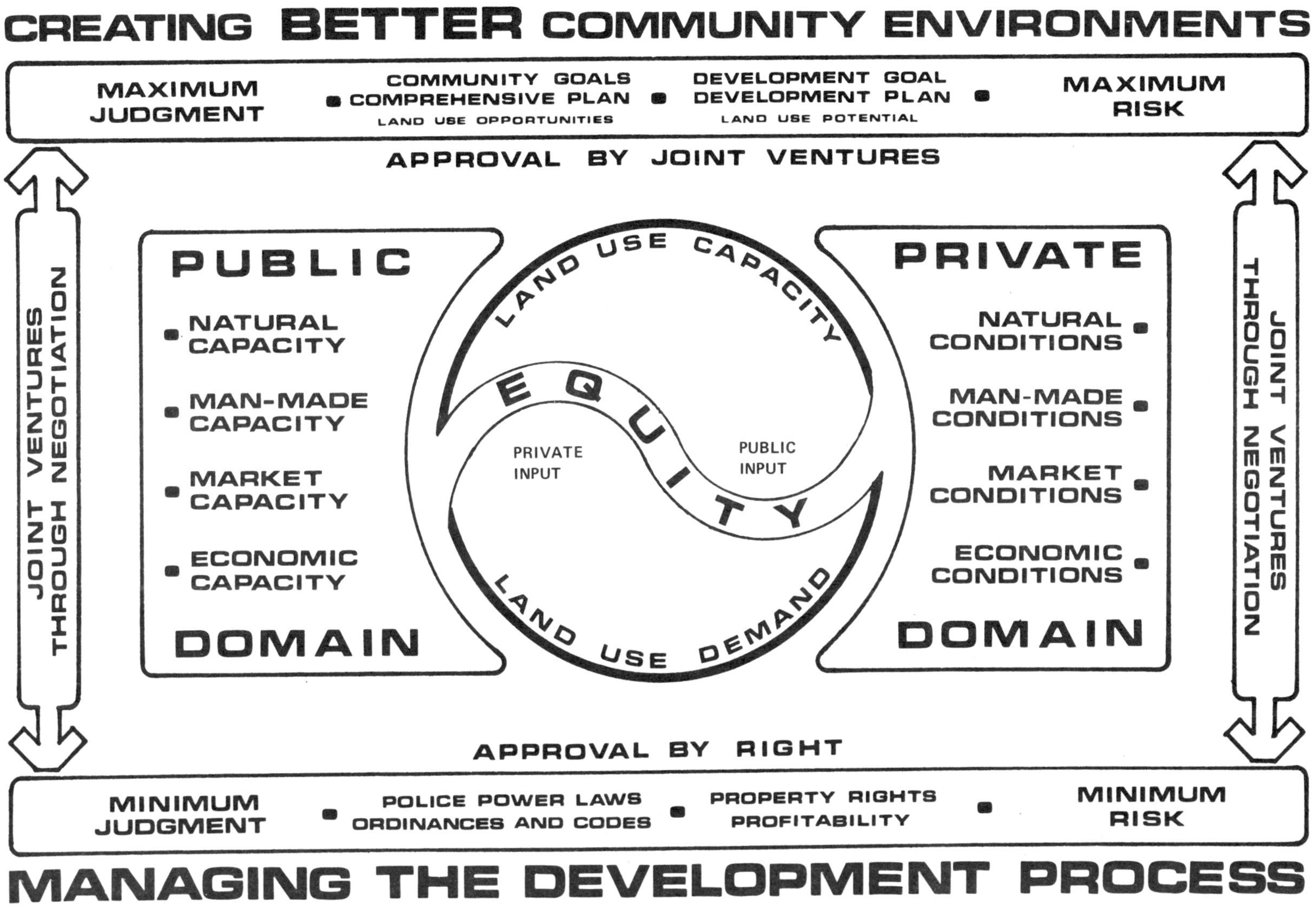

FIGURE 3 Managing the community development process through public and private joint-venturing.

obstacles to improved productivity. Of the many causes that have been advanced for our present "failure" to solve problems--apathy, carelessness, negligence, greed, and ignorance--by far the most serious is ignorance. Our collective ignorance or, more simply, our lack of knowledge plagues us most. We need to take better advantage of our local idea resources and create an improved local information base for more productive decision-making. I am arguing for the rediscovery and, if necessary, the rededication of the educational resources and facilities available in every local community. My experience shows that systematic instruction using the community as a classroom at the elementary, secondary, and adult education levels produces excellent new ideas, provides new information required to complete the local information base, and improves leadership skills of all involved. Young citizens should be learning about community problems and the means to solve them in their school classrooms rather than waiting until they are elected or appointed to office. Ideally, one who stands for election, one who agrees to serve on a community commission or board, and all administrative staff should accept, as a condition of serving, the continuation of their education in community problem-solving in a productive, formalized way.

Most community problems are people problems. They become problems as people pursue their own interests. To find solutions requires designing new arrangements and new delivery systems that bring the enlightened self-interest of both public and private decision-makers to bear on problem.

Our present decision-making and actions, good or bad, create a legency for our children and grandchildren. Creating and enabling more preferable futures for tomorrow is the responsibility of us all.

DISCUSSION

QUESTION: What are the building trade unions doing to recruit and train females so that union contractors will have sufficient women to meet the Department of Labor female hiring goals and timetables? In addition, how do you think the presence of females on the job site will affect morale and productivity?

MR. GEORGINE: The Building and Construction Trades Department of the AFL/CIO presently is not recruiting women on a large scale. Women are not knocking down the doors of the union offices asking to be trained as apprentices in our various crafts. We do have some women, but we will never meet the quotas because the quotas and the goals are unrealistic. One reason for the productivity problem in the construction industry is the strong emphasis that has been placed in recent years on training for social reasons instead of training to provide functional journeymen for the building process. In my opinion, the Comprehensive Employment Training Act does not belong in a construction industry. The goal of construction training is not social reform. Our aim is to train people to perform certain tasks and functions in the building process. We must go back to training people to do a job properly and not just because they are white, black, female, or whatever.

MR. McARTHUR: As far as motivation is concerned, we at the Austin Company have not found that the presence of women affected the job. They were a novelty at first, but then everyone got used to the situation. Actually, our biggest problem is motivating females to stay on the job. The drop-out rate is 10 to 1. They come on the job because they have heard about the big money, but construction work is difficult at best and, as soon as they find that out, they leave. We are dealing with a different kind of minority problem here. Many minority workers are used to hard work, to moving around, and to getting laid off. Females, however, have not had that kind of experience. We try to motivate them to stay, but it just is not working.

QUESTION: Is there any way to reform the troublemakers that are on the job? What are the unions doing to help reform them? If you fire them one day, will the union refer them back the next day?

MR. KELLSTROM: We had a very serious problem with such people in Florida. We expect our supervision to be both fair and firm. What happened was that about 80 electrical workers had a meeting. We did not know what was being promoted, but we did know that the men were not working. Our supervision was firm enough and fired all 80 men. We would not have done that unless we had the support of the trade union and, in this care, the union business manager supported us wholeheartedly. We did not get one of those men back. I do not know what happened to them, but I hope the union put them in different areas because en mass, they are terrible.

MR. GEORGINE: The unions support the contractor in those cases when it actually is shown that a disruptive group is on the job and that its only purpose for being on the job is disruption. We try to cooperate every way we can, within the law, with the contractor in his attempts to eliminate that kind of a problem.

QUESTION: If unions are interested in improving production, why do they resist featherbedding reductions and production-improving mechanical devices?

MR. GEORGINE: The union's function is to negotiate, for the people it represents, working conditions that are not injurious to health and good wages. Obtaining more productivity is really a management function, and the unions generally do not resist the use of new technology. The question, for example, of whether or not an oiler is needed on a hoisting rig can be debated. There are functions that oilers should perform all the time; however, if there is something in the collective bargaining agreement that prohibits him or restricts him from being used productively all the time, it should be taken out of the collective bargaining agreement through the collective bargaining process. I think that much more is made out of this issue than really exists, but I do think it is a collective bargaining problem. It became a part of an agreement because management agreed, and management should use the same process to eliminate it.

MR. McARTHUR: Some problems have occurred because of changes in technology and equipment. Oilers, for example, perform a very important function on the older machines, but now we also have 60-ton cherry-pickers with hydraulic outriggers that are one-man operations. Thus, the oiler does have something to do on some types of crane and absolutely nothing to do on other types. Negotiation is required. Management has to sit down with labor and negotiate the restrictive clauses out of our agreements. In the past three or four years, this has been done effectively in some parts of the country but not in others. It is nevertheless a labor-management negotiating problem.

QUESTION: Is it true that time and motion study surveys are illegal on government funded projects?

MR. CURTIN: I would be surprised if they were, but I do not know the answer for sure and cannot give you a legal opinion.

QUESTION: It seems that the relative impact of union featherbedding, work rules, and jurisdictional disputes on productivity growth has not been examined or debated. Is this because the general consensus of management is that such things are relatively unimportant? In addition, am I correct in assuming that the growth of the merit shop in the past 15 years has stimulated the union trades council to change its attitude toward jurisdictional disputes, work stoppages, featherbedding, and work rules?

MR. GEORGINE: I think you raise some very valid points that are worthy of discussion. One of the problems with measuring productivity in construction is that what we use to measure it is so vague and obscure in everyone's mind that we are not always measuring apples for apples and oranges for oranges. When you speak of the merit shop and its effect on productivity, you really are asking whether the merit shop system is more productive than the collective bargaining system. There are very few statistics on which to base such a comparison. We know that heavy industrial work is accomplished in fewer man-hours when union forces are used under the terms of a collective bargaining agreement than when non-union forces are used. However, the mix of men that are paid for those man-hours is different, and the cost aspect is therefore different. Thus, more man-hours may have been used but the final cost might be lower.

We simply do not have the data we need and it is very difficult to get them. The kind of figures that are given on open-shop jobs are not the same kind of figures that are given on union jobs. For instance, we know that there is much more supervision on a non-union job than there is on a union job; therefore, that cost is not reflected in the worker cost as it is in the union job. How can we get the kind of data needed for a comparison? We would need to know how many man-hours were actually spent on a job, the ratio of supervisors to workers, the pay of the various workers, and many other things, but we cannot get that information. Thus, most statements on the subject are not valid because different things are being compared.

It may well be that the growth of the open shop over the past 15 years has accelerated the removal of restrictive clauses from collective bargaining agreements; however, the kinds of facts and figures available on jobs done on an open-shop basis may not be accurate. I would welcome a valid comparison if we could get adequate and compatible data. If you compare, for instance, a heavy industrial job done today with one that was done 15 years ago, you will find that productivity is much higher today. But why? Not so much because of technological changes, but rather because of how manpower is used. Fifteen years ago, the user or owner of a heavy industrial project would demand that the job be done by a certain day. The contractor, who did not have a hard money job, would put more men on to satisfy the owner. The contractor did not care if the workers were being used properly; he was just satisfying the owner. Suddenly, someone looked at the cost of construction, and the owner and the contractor had to figure out how to reduce costs. They decided to get rid of unions and their restrictive provisions so they brought in the open shop. They never asked the unions and the workers how to reduce the cost of construction. If they had, we would have told them to eliminate half the workers and supervisors who do not know what they are doing. But we were never asked, and that is part of the problem we are discussing today. We now are trying to get involved in the process so that we can present our ideas on how to cut the cost of a job. There are a lot of evils in the construction industry today, but there also are a lot of goods. The problem is to sort them out, and the only way we can do that is to try to find a solution together.

MR. KELLSTROM: With respect to the question about featherbedding, I contend and always have contended that, although we always will have our share of featherbedding, we will never quit trying to do something about it. And I blame management, not unions, for featherbedding. You can do something about it, and the quicker you do it, the better off the construction industry will be.

MR. McARTHUR: I have one comment about work stoppages over jurisdictional disputes, and here again, management is at fault. The secret of good jurisdiction is to know what you are doing, and a lot of companies do not. Jurisdiction with my company is really a minor part of the problem. When we have a big job, we do what we call a "mark up" prior to ever starting the work. We go through the prints and we identify every piece of equipment, everything that is going into the job. We lay it all out and note how we are going to assign the work. Then we sit down with all of the trades and go through this document. If there is a dispute on some work, we so note it. If the two unions cannot agree that day, we give them a day to resolve it. That document becomes a permanent part of the job, and everybody has a copy. If you do not plan your jurisdictions, you are going to have trouble.

MR. GEORGINE: Many people do not realize that there are occasions when labor problems on a job are created for the sole purpose of having a reason for not meeting a deadline. Many contractors are good salesmen, and when they sell the job to the owner, they specify certain completion dates. If they later find that they are not going to meet that schedule, for whatever reason, and there is some kind of penalty clause, one of the simplest things to do is to make a bad assignment that creates a labor problem, possibly a walk off. They then can go to the owner and say they are behind because of labor problems. That may seem devious, but it has happened.

We have to look at the whole productivity problem in a more positive way. We have to develop measuring sticks for determining what productivity really is in the construction industry. There have been many productivity commissions but I have yet to find some way of accurately measuring productivity on a construction job. Most of the measurements are visual, and as you watch a construction job, you inevitably see people standing around. There are reasons why those people are standing around, but you do not know what the reasons are and only assume that you are paying someone to waste time. If there is no valid reason, those people ought to be fired and the unions will support that kind of action by a contractor. Another one of the keys to good productivity is to have economic conditions that permit continuous work so that unemployment does not occur. Rebuilding and retraining a workforce is very costly in terms of productivity.

Another key involves the regulatory system. On some jobs (especially nuclear power plants), there are more inspectors looking at everything that is done than there are workers. They impede progress if the workers must wait for an inspection for a long time before they can proceed. In addition, having supervisors and regulatory personnel standing around talking and having a good time is not very good motivation for the workers. These are some of the human things that affect productivity. I do not know if there are solutions to these kinds of problems, but they do affect productivity.

MR. SABGHIR: I am delighted that this discussion has gotten us to the place where we can talk about some action that may result from this particular conference. It has troubled me that maybe we will not get any action, and I now would like to make a suggestion. The past few minutes resulted in the surfacing of an issue that normally does not get the kind of attention it deserves--i.e., information, knowledge, facts. I have spent a good part of my life arguing and fighting for personnel and being only partially successful in trying to develop better information about the construction industry. To make a long story short, I tried first to get the federal government to do the right thing with only partial success. I have tried to get the industry to organize itself for better data and, again, with only partial success. Seven or eight years ago, I attended an interagency meeting made up of representatives of different agencies, including the Department of Labor, and one of the issues I raised was the inadequacy of information concerning the nonunion sector of construction. I will never forget the response of the Labor Department spokesman who, in effect, said that it is too hot a subject to discuss. And now this question surfaces here. I also remember the 1960s when we thought we had inflation and President Johnson established a labor-management committee. I briefed the Secretary of Commerce for one meeting on inflation. I said the big problem was that labor and management people will invariably argue about the facts. Labor will say they do not agree with the facts, and management will say they do not agree with the facts. I was right; the meeting broke up over the issue of what the facts are.

We have arrived at the time when labor and management should get together, with or without the government, and try to agree on the kinds of information that we need about the construction industry (and not just from the point of view of productivity). If the labor forces and the management forces can get together as one single voice, the Department of Labor will not have to collect labor and material use surveys using 20 projects, for example, in order to represent all of the projects in the United States. We can do a much better job. It will take money and money is always hard to get, but if labor and management get together, we at least can concentrate on something that could have enormous potential for use in analyzing the problems and in identifying where we want to go. I would propose that this Forum will be very successful if it generates nothing more than a joint labor-management endeavor to get together and agree on what facts are needed.

MR. SCHMITT: It bothers me that this discussion has focused almost exclusively on the productivity problem with respect to the union-management question. It is a valid part of the issue, but the regulatory problem is even more important. We have discussed union work rules and so forth at length, but we have not mentioned that there is more inflexibility in our regulations than in union work rules. Unions change, but a law, once on the books, hardly ever changes. Plumbers, by and large, are not nearly as inflexible as plumbing codes. Carpenters, in particular, are not nearly as inflexible as the building codes that regulate the use of wood. Electricians are not nearly as inflexible as the codes that are written to regulate the use of electrical devices. Everybody seems to be missing this point. Our real productivity problem does not stem from unions and management. Our problem is regulation that stops the use of technology and sterilizes the management of the industry.

QUESTION: What incentives are there for the construction industry to train for and to invest in technology for conversion to the metric system for measurement? Will productivity be affected?

MR. McARTHUR: Conversion to metric is not a serious problem. We do a lot of overseas work, but most of our design work is done in this country so we use conversion calculators. Almost all of our overseas work is engineered, designed, and shipped out of the country in metric form. It has not been a serious problem.

Session V

INNOVATION FOR IMPROVED PRODUCTIVITY

PANEL OVERVIEW

JEFFREY J. HALLETT (Moderator)
Director, Productivity Center, Economic Division
Chamber of Commerce of the United States, Washington, D. C.

This panel is composed of a unique set of individuals, and it represents about as good a collection of expertise and experience in the field of productivity improvement as could be assembled. The panelists come from several countries and represent organizations specifically directed toward productivity improvement. In addition, they reflect construction field experience at all levels of size and complexity.

To begin, I will present an analogy that has to do with the continuing discussion of productivity, its measurement, and innovation. Assume that you are in a laboratory and want to conduct an experiment about innovation and productivity using white rats. The object of the experiment is to determine what training or conditioning will cause a rat to become most adept at finding cheese at the end of a maze. You have a maze with five tunnels and you put a piece of cheese at the end of the fourth tunnel. Then you drop a rat who has never seen the tunnels into the maze. He scurries up one tunnel and does not find the cheese. Then he scurries up another tunnel and does not find the cheese. Soon, however, he comes to the fourth tunnel and he gets the cheese. We then replace the cheese and start again. He scurries up the tunnels looking for the cheese at the end of the fourth tunnel, and the more he does this, the better he gets. In fact, we find that each time we put the rat into the maze, he gets better and better at getting the cheese at the end of the fourth tunnel, and, soon, he runs down that fourth tunnel immediately and gets the cheese. Our very productive rat has a problem, however, when we move the cheese to a different tunnel. He runs down the fourth tunnel, but does not find the cheese; nevertheless, he keeps running down that fourth tunnel looking for the cheese until he drops. On the other hand, if we put an untrained rat in the maze, he will go down the first tunnel, and if the cheese is not there, he will go scurrying down the other tunnels until he finds it.

In many instances, that has been the way we have measured productivity. We have sorted out the way we do things, and we have gotten very good at getting down the fourth tunnel. However, one of the things that has happened to our industry and to the economy in general is that conditions are such that the cheese keeps getting moved. Therefore, it is incumbent upon us to continue to find ways of getting the cheese without getting too attached to a particular tunnel.

PRESENTATIONS

DAVID S. MILLER
President
David S. Miller & Associates, Inc., Cleveland, Ohio

Most of my presentation will reflect my work with the National Institute of Building Sciences; however, I also will take the liberty of commenting from my own personal experience in areas that I think are of special interest with respect to the improvement of productivity. First, let me note that the Forum program states that innovation will come about from "the sharing of ideas, technology transfer programs, and an evolution from traditional operational philosophy." I would like to stress the last of those three points. It already has been said many times that we too often are bound by traditional philosophy even though we all want to get away from it.

In this context, I would like to point out that the National Institute of Building Sciences (NIBS) is something new and unique. For the first time, Congress proposed, through legislation, the creation of a private, nonprofit corporation that was to be essentially catalytic in nature and that also was to be bound by mandates and legislation to utilize, to the best of its ability, presently existing organizations and to encourage them to work together more effectively. It is very significant that many segments of the construction industry, an industry too often described by the words "traditional" and "lack of change," are asking for change, and now I believe we have the opportunity to accommodate and create that change.

The Forum program also explains that "improved productivity is a major factor in accomplishing results from technology change and that it can be obtained when the building and construction community unifies in a national effort to reach solutions to problems." That simply underlines what so many have said. Let us finish this conference and then be sure we do something about it.

The legislation that created NIBS, in effect, established the objective of encouraging methods that would improve the regulatory process and structure. Another basic goal is to encourage the removal of constraints to new technology and innovation. In effect, we must accomplish both to make a major contribution to improved productivity.

As part of its efforts in this direction, NIBS recently completed a study of constraints on innovations and new technology. Figure 1 summarizes this study and illustrates how far-reaching is the question of constraint on innovation and technology with respect to the construction industry.

The building community too often is thought of as being either technologically strong or technologically weak, depending on whom you are talking to, but most often technology change relates to the producer who has a research and development (R&D) capability and therefore is supposed to produce something new and innovative. But when you examine all aspects, the question of constraints on innovation and new technology goes across the board.

The members of the building community covered by this study include the owner-user sector, the financial sector, the manufacturer-supplier sector, the design sector, the construction sector, the labor sector, the regulatory and code sector, the information and education sector, and the research and development sector. To illustrate, I will mention only a few of the factors that were identified in this study as having an effect on productivity.

For example, with respect to the regulatory and code sector, Figure 1 refers to "existing problems which provide the stimulus for innovation." These problems sometimes are looked on as negatives, but, in this case, they are considered to be positive in nature. Included is the proliferation of regulations as well as the lack of a standard terminology and a standard interpretation of regulations, which results in an increased burden of regulations in all sectors.

The next column in Figure 1 identifies the influence of these problems on productivity. For example, they cause uncertainty in all sectors, and both inhibit and stimulate some design. The next column identifies the constraints to improving this situation--i.e., the system of regulations is cumbersome, and there is a lack of uniform interpretation. Thousands of local area interpretive bodies are developing individual judgments on code matters. What, then, are some of the possible solutions? Coordination of various agencies and code groups is an important solution, and much already has been accomplished. It might be worthwhile to take a look at the existing codes and regulations, to relate them to rehabilitation of old buildings, and to develop new guidelines, looking also at mobile home studies. We then could try another approach by looking at them a little differently. Add in the effort to encourage the development of performance standards and codes with new measures of accomplishment.

In the manufacturer-supplier sector, some of the actions that provide the stimulus to innovation are changes in consumer demands, a change in supply of resources, and a desire for market growth and profit. What is their influence on productivity? Well, they stimulate development of new products and specifications and encourage manufacturers who desire market growth to push new products through the system. What are the constraints to these improvements? There is the reluctance to invest in R&D due to the uncertainty that comes from a proliferation of new regulations and the negative effect on future economies. What can be done about it as possible solutions? Encourage the development of a performance approach, with its consequent option for developing different solutions, or improve, simplify, or remove constraining regulations. There are many more items listed in the matrix (Figure 1) that are significant in improving opportunities for innovation.

EXISTING PROBLEMS WHICH PROVIDE STIMULUS FOR INNOVATION:	THEIR INFLUENCE ON PRODUCTIVITY:	CONSTRAINTS TO IMPROVEMENT:	POSSIBLE SOLUTIONS:
Consumer Sector: o Tighter budget, greater selectivity o Change in program requirements - energy budgets, life cycle cost - barrier-free design	o Fewer building starts o More competitive designs	o Economic trends which cause the problems are continuing	o Consumer education to ensure appropriate selection of designer, materials, program, etc.
Financial Sector: o Increased concern for security of "risk" o Higher interest rates o Threat of reg. sanctions thru FDIC lenders	o Increases the time required to start projects o Often penalizes innovative approaches	o Supply and demand o Lack of thorough understanding of the benefits of new concepts	o Education: - lower operating cost increases borrowers cash flow - better risk to lender
Manufacturers & Suppliers: o Change in consumer demands o Change in supply of resources o Desire for market growth & profit	o Stimulates development of new products and applications o Prompts manufacturers to "push" new products thru the system	o Reticence to invest in R&D due to uncertainty of future regulations & economic markets	o Encourage development of "performance approach"
Design Sector: o More competition for fewer jobs o Change in consumer demands o Increasing liability o Desire to innovate	o Increase the need for creative solutions o Causes design professionals to forego profits in exchange for less liability - diffusion of responsibilities	o Uncertainty, fear of increased liability o Difficulty of obtaining appropriate information	o Encourage creative solutions to problems of liability, certification of compliance o Tailor information & education to user needs
Construction Sector: o Increasing cost of labor and materials o Decreasing on-site quality control o Labor disputes, delays, etc. o Increasing liability	o Tends to preclude use of innovative approaches - perpetuates traditional methods and materials	o Fear of delays, etc., raise bids for completion of innovative projects	o Encourage coordinated construction management o Training programs to assist introduction of new methods and systems
Labor: o On-site conflicts with prefab units o Division of labor conflicts over modular systems o Increasing cost o Lower productivity, less skilled labor available	o Prompts designers to circumvent the system prefabrication, components, etc.	o Preconception that increased productivity means fewer jobs	o Coordinate labor unions on- and off-site to facilitate uniform acceptance of manufactured components, practices, etc. o Encourage on-site quality as means to maintain job security
Regulatory & Code Sector: o Proliferation of regs: BEPS, OSHA, EPA, etc. o Lack of standard terminology & interpretation o Increase burden on all sectors	o Causes uncertainty to all sectors: - stimulates some (design) - inhibits most	o Cumbersom system o Lack of uniform interpretation	o Coordination of various agencies & code groups o Cull existing codes & regs (Rehab. guidelines, Mobile Homes study, etc.) o Encourage "performance approach" with standard measures of compliance
Information & Education Sector: o Need for information & education in areas of energy conservation, management techniques, new methods and materials	o Stimulates investigation & education o Prepares professionals for changes in in other sectors	o Too far removed from the "real world" o Information is not always available in the form or at the time it is needed	o Coordination of producers & users of information
Research & Development Sector: o Duplicated , missed, not coordinated o Results not disseminated	o Slows the introduction of new technology	o There is little coordination of R&D efforts	o Coordinate R&D efforts

FIGURE 1 **Summary of constraints to innovation in the construction industry (from a National Institute of Building Sciences staff report).**

I would like to turn now to a different topic, a specific area that I think can give us all a little pause for thought because sometimes we can overlook productivity opportunities that occur in places other than the actual building itself. This opportunity for productivity gain relates to office work. Because of our involvement in the construction process, we naturally look at the construction of a building as the end result. But, in truth, if we think not only of the building itself but also of the building as housing an economic enterprise for some 40 years, the normal life of that kind of a building, we will look at that building a little differently. It is no longer just a shell, good looking as it may be: it is housing a business that is going to continue for 40 years or more. Based on these assumptions, a special study was conducted some time ago. Using a typical office building and looking at it as housing an economic venture for 40 years, it was concluded that the initial building cost represents only about 2 percent of the total 40-year cost. Maintenance of a type to keep that building operating in normal ways represents about 6 percent of the cost over the 40-year life of that building, but the total "people cost" to make the economic venture function is 92 percent. All of a sudden we have a little different kind of potential target when it comes to the opportunity for productivity gain.

Now we must add another set of factors. Work changes in an office building. Thanks to the American Telephone and Telegraphs of the world and many of their friends and competitors, there have been many developments in the electronic area, including new forms of data communication and telecommunications. At the same time, traditional types of office work have been going on in the office building. Now, the new electronic equipment and the typical office worker are finding a new relationship. The data communication industry expects to demonstrate productivity in the use of its equipment by putting its equipment where people work, but to put a piece of equipment on a desk does not necessarily mean that the expected productivity will result. The physical environment of that office becomes very important.

Mr. Georgine made a very true statement when he said that productivity is a very human kind of thing. How can you expect this wonderful machinery, which is designed to produce more results, to be properly used from a productivity standpoint unless the people who are using it are in an atmosphere and environment that respects them as human beings? This represents a completely new relationship to be considered with respect to the product of our industry, the buildings where people work. Too often, however, these new office situations are considered in a productivity sense in the same way as on-the-job productivity or construction process productivity.

I suggest that these are important areas relating to innovation and new technology. There are great opportunities for improving productivity that stem from innovation and technology, but we must seek them in a new way.

Currently, a study that relates directly to this office productivity opportunity is being conducted by the BOSTI organization under a grant from the National Science Foundation (NSF). The initial NSF grant triggered other private funding, and the study now is moving into an analysis of the human environment, how work is accomplished in it, and how greater productivity can be accomplished. One important aspect of this study is that it will analyze the situation in 10,000 or more workplaces where typical office work is done. Half of the office workers in the study are in the government sector and half, in the private sector. The sponsorship of the study involves both private sector and government organizations. The results of this study can add tremendously to the information we need if we are going to find practical and pragmatic answers to productivity questions.

In conclusion, let me repeat what I stated earlier. Nothing is more important with respect to accomplishing productivity gains than to bring industry leadership together on this subject. That is what has happened here. Let us carry it forward. What we do when we leave this meeting probably will mean a great deal more than what we have done while we have been here. I for one hope that we all can do something about accomplishing real productivity improvement.

SEYMOUR BORTZ
Senior Engineering Advisor, Materials Technology Division
Construction Productivity Center
Illinois Institute of Technology, Chicago

One of the initial objectives of the Construction Productivity Center was to determine what the industry thought about productivity and how it could be improved. As researchers, we felt we could not just jump in based on our own experience or what we thought improved productivity should involve. Therefore, we sent a questionnaire to Engineering News Record's 400 contractors, and my presentation will focus on the responses to that questionnaire. These responses are important because they give us some idea as to what the industry itself thinks is important and how it thinks increased productivity can be obtained. As earlier speakers have said, the management function is very important, and it is interesting to note that those responding to our questionnaire generally held very high positions. In addition, the response to our questionnaire was good indicating, I believe, that many construction companies think improved productivity is going to be helpful to them and that they are interested in working toward it. Since time will not permit me to present a detailed description of the questionnaire results, I have prepared a statement that reviews some of the basic facts concerning the construction industry and summarizes the results of our questionnaire.

The construction industry is the largest industry in the United States. It employs over 10 percent of the work force and contributes over 10 percent, or more than $100 billion, of the country's gross national product. The industry is highly fragmented and diversified, and it is composed of over 800,000 construction contractors, ranging in size from a few giants who employ thousands of people to the majority who have fewer than 10 employees.

Research and development is almost nonexistent in the construction industry. It has been reported that the construction industry spends only a fraction of 1 percent of the industry's gross revenue for applied research as compared to the relatively small electronic industry that invests between 10 and 20 percent of its revenue for research and development.

The United States is facing an economic crisis: inflation, high unemployment, and the declining value of the dollar. At least one of the factors contributing to this crisis is the lack of productivity improvement in the U.S. economy over the past decade or more.

In attempting to maximize productivity improvement in construction, it is important to focus on those areas that represent the greatest potential for productivity improvement rather than to spread our efforts equally on all areas. In other words, it is important to identify and attack the vital few areas as opposed to all areas at once.

The major problem is pinpointing those areas with the greatest potential. It would be naive for a handful of individuals to identify areas for the entire industry. For this reason, we sent out our questionnaire to executives in the construction industry. We believed that their replies, based on their knowledge and experience, would be invaluable and any conclusions drawn would find ready acceptance since they were based on information from the industry itself.

Out of the 400 questionnaires distributed, 99 or slightly less than 25 percent were returned. The usual return on any kind of a questionnaire is only 4 or 5 percent; therefore, this high level of response is significant in that it indicates that construction executives feel this subject is important. Fifty-nine percent of the questionnaires were answered by either the company president or chairman of the board, and 28 percent were answered by vice presidents. Thus, this survey could be called "A Construction Executive Survey for Productivity Improvement."

To analyze the results, we grouped the companies according to size and categorized their responses about the potential for productivity improvement in various areas as high, medium, or low. For simplicity, the contractors with greater than $500 million annual sales were designated as Group A; $100 to $500 million, Group B; $50 to $100 million, Group C; and $10 to $50 million, Group D. Obviously, we are talking about relatively large companies.

The first series of questions focused on headquarter's opportunities for improvement. All groups felt that planning and scheduling offered a medium to high opportunity for improved productivity. Group A indicated a low opportunity in the estimating function, whereas the other three groups indicated a medium to high opportunity. All of the groups indicated a medium to low opportunity for improved productivity through specification improvement. All of the groups indicated a medium opportunity for improved engineering.

The second series of questions addressed construction site opportunities for improvement in various functions. Under the management function, Group A indicated a medium opportunity for improving productivity through supervision, whereas the other three groups indicated a high opportunity. One could interpret this to indicate that very large contractors believe their supervision is adequate and that the other contractors believe a major improvement is possible.

Under the engineering function, the Group A contractors indicated a low opportunity for productivity improvement by improving design standards and standardizing specifications, whereas the other three groups indicated a medium to high potential. In design improvement and systems engineering, Group A indicated a medium opportunity for improvement and the other three groups gave this category a rating from medium to high. With respect to construction techniques (precast, preassembled elements, and other types of developments), all groups indicated only a medium opportunity for improving productivity. In other words, they feel that they can get some additional improvement in productivity through the use of preassembled and precast elements but that they do not rate the potential as extremely high.

In connection with improving regulations to improve productivity, the smaller contractors, Group B, indicated a high opportunity. The other groups were divided between low and high, so there was quite a discrepancy in their ideas on that.

Under the labor function, contract agreement received a rating from medium to high. It should be noted, however, that several personal interviews were conducted in connection with the mailed questionnaire and that in the interviews the contractors indicated a high opportunity for improving productivity by improving labor contract agreements. When the contractors answered the questionnaire in private it seems that their attitudes mellowed.

The responses with respect to labor training were another surprise. The very large contractors, Group A, indicated predominantly a low opportunity for improved productivity with improved labor training. The other groups indicated a predominantly higher rating.

The responses about the final function, equipment, also were something of a surprise. Improvement in the capacity, simplicity, and maintainability of construction equipment has made great strides over the past decade. However, all contractor groups strongly indicated a medium opportunity for further improvement.

The final series of questions concerned contractor participation in improving productivity and was of particular interest. The responses generally indicated a strong willingness by the contractors to contribute something to improve productivity. This is what we really are talking about here at this Forum--getting the whole group of contractors involved in doing something to improve their own position--and it appears that they are willing to do so.

The answers to the questions were quite similar from the four groups. Over 40 percent of the contractors indicated they would serve as a member of a group to identify productivity problems (so that will help answer and quantify how you measure productivity), conduct productivity conferences, and subscribe to information services devoted to construction productivity.

Over 25 percent of the contractors said they would contribute funds and develop projects aimed primarily at improving productivity. Again, they are willing to cooperate and see if they can quantify productivity in such a manner that it can be measured and we can show improvement.

In comparing the four groups, it would appear that Groups B and C were more willing to develop projects and contribute funds than Groups A and D. This was a rather interesting alignment since Group A represents the more affluent companies and Group D, the smaller contractors.

As a result of this analysis, it became obvious that the smaller contractors had a definite viewpoint that really differed from some of the larger contractors. Many of the contractors who participated in the survey not only took the time to answer the questionnaire but also provided additional comments relating to the topic. Some of these ideas have been brought out in our discussions here. The comments included the following:

- If management does its job in organizing, planning and supervision, the result cannot help but improve productivity and job satisfaction.
- A prerequisite for productivity improvement is long- and short-range planning and scheduling, strong material expediting and control, knowledgeable and cost-conscious superintendents, and good cost control and unit measurement systems. Again, we are talking about how you measure the productivity.
- People of the United States have lost the work ethic that built this country. Management's inability or unwillingness to cope with problems, government regulations that slow down capital investments, the decline in research and development, and management's reluctance to participate in trade associations are the major problems in construction productivity. The "why should I care" attitude is too predominant.
- Eliminate training restrictions in the Davis-Bacon Act, legalize non-union training programs, negotiate project agreements to eliminate featherbedding, negotiate special agreements like those for residential construction, and provide for compulsory arbitration.
- Government regulations slow down productivity. Labor is losing its desire to be good craftsmen. Think United States and not foreign development. The people's attitude is the biggest problem.

- o Get rid of the Bacon-Davis Act, and you will get rid of 50 percent of the productivity problem.

You can see that there is great interest on the part of contractors to see what can be done with productivity improvements, and I believe a large percentage of them are willing to put their money where their mouths are. I think it is up to those of us attending this conference to provide them with the forms necessary to define just what construction productivity is and how it can be measured and then to get ourselves involved in what can be done to improve it.

JOHN W. LEONARD
Vice President
Morrison-Knudsen Company, Inc., Boise, Idaho

Morrison-Knudsen is involved in many activities but I will try to concentrate on the construction portion of its business. Speaking as a contractor, then, I will first note that ours is an undercapitalized industry. This has been true for some time. The trend began in the early 1950s, but the current high inflation rate is taking its toll and making the replacement of equipment more difficult. If you replace equipment at today's high costs, you eat up cash, but if you hang on to it, you run the risk of high repair costs and considerable down time. The so-called high interest rate, which now has become almost normal, also has an extraordinary effect on the industry's capital. Thus, the potential for using equipment to improve productivity is decreasing. Fundamentally, the construction business needs higher profit margins. That we do not have them is partly our fault and partly the result of owners' buying and contracting practices. And you generally get what you pay for.

My second point concerns disputes. What contractors basically do is construct large projects outdoors, exposed to elements, using temporary facilities in contact with the earth with its awesome scope of varying conditions, many of which cannot be predetermined economically. This leads to disputes when changed conditions are encountered and disputes lead to lawyers. Note, first, that there are no price-wage guidelines for lawyers and, second, that disputes also result in delays. Since we cannot simply eliminate all lawyers, I suggest we consider arbitration, mediation-arbitration, and contract appeals boards to lessen the impact of the strength-sapping dispute resolution process.

I also believe that we contractors should stop complaining about regulations and unions. Granted, both may cost us something, but the union is our supplier of a most valuable resource--generally qualified people to be lead by us.

I do not have any trouble measuring productivity; it is cost. Our figures show that heavy construction productivity in particular improved from 1945 to 1960. The decline started slowly in the early 1960s, although it was disguised by vastly improved equipment. This disguise disappeared when prices increased and equipment improvements became fewer in the late 1960s and early 1970s.

One of the reasons for this decline might be that good supervision started to disappear in the 1960s. Many foreman did not want to leave the ranks. Year-round pay of the superintendent did not mean that much. Wages had gone up and employee's pensions, unemployment pay, unemployment insurance, and income tax all entered into it. The supply of those who liked the taste of leadership they got during World War II declined. The industry then tried to rely on university-educated engineers to make up the difference. We also tried to rely on new glamor things such as planning, scheduling, and computers and put them in unproven hands. Only now are we straightening out this situation. Let me give you an example in terms of planning and scheduling. The only significant source of good planning and good scheduling is a good estimator. He has to schedule properly to price properly, and he has to recognize the effect of each element on a project. Estimators take a final exam of sorts every few weeks when bids open. The poor ones go. I believe Mr. Bortz indicated that his Group A did not see much room for improvement in their estimating and bidding process but that they saw a lot of room for improvement in the scheduling process. That is true of many companies, and there is a message there.

Until people who understand and are experienced in planning and scheduling are well integrated in the process, productivity will decrease as we try to meet unrealistic and unsound schedules with a poor plan. For several years Morrison-Knudsen has carried on in-house training programs for supervision and managers to overcome this problem. But good supervision is like a good woman--hard to find. There are not many and the difference between good and not so good is about plus or minus 15 percent, a total spread of 30 percent. We must develop more and we are trying.

Until the industry develops enough good supervision and management, I think you should consider a few things. It is impractical to have projects with more than 2500 to 3500 workers. You should insist on a minimum of 30 percent company hands on the project up to a maximum of 75 percent. A company hand is somebody who has been with the company for three to five years. You are going to have to pay these people. An earlier speaker alluded to this cost as 2 percent of the total, and I think we have been buying the cheap quality in a lot of cases.

Finally, if a contractor develops something innovative, it has cost him time, effort, and money, but he can only reap a short-term benefit since everybody in the industry gets to know everything in a short time. This short term is also generally the agony term when he is still trying the improvement. Once it is proven, everybody gets it at the developer's expense. Surely, the contractor could patent or copyright his innovation, but how does he market his innovative idea to get some benefit, to recoup his investment. This should suggest something. Could not one or several existing organizations take on the role of using royalties to pay for the expense of patenting, copyrighting, and marketing?

To summarize, let me repeat four major points. The construction industry needs to make a bigger profit. We sell too cheap or the owners buy too cheap. Second, steps are needed to resolve contractual disputes quickly. Third, more in-house training is needed to develop good supervision. Fourth, possibly a new innovative mechanism should be developed. Productivity and innovation will result.

R. E. JEANES
Head, Planning and Construction Department
Building Research Establishment, Garston, Watford, England

Industries in the United Kingdom and the United States are different from one another in many respects--not least important is the influence of government as a client and the different contractual boundary between the architect and the contractor as I understand it. I will try, however, to describe the general lines of our research program because they are felt to be of some importance in the general search for greater efficiency in building and I will then present a specific suggestion derived from our studies over the years that seems to be applicable at whatever level of productivity one is operating.

The concern of this Forum for all aspects of the building process is one that would be echoed in the United Kingdom. My own organization, the government-funded Building Research Establishment, deploys about 50 staff on research studies seeking to establish ways and means of improving overall efficiency in the process.

Our current program includes studies with the following objectives:

1. To classify the effects on firms of changes in the level and pattern of demand so that better understanding will lead, at best, to government action to minimize the effect or, at least, to a better basis for industry planning. This program is designed to establish better facts. We attempt to establish the labor and materials input that will be required by a particular demand pattern upon industry. The demand pattern in the United Kingdom is changing from new building to rehabilitation and this trend calls for a quite different pattern of labor requirements.

2. To improve the process of briefing and initial design including improved cost advice. I have been much impressed during this Forum by the extent to which the owner of the building figures as an important contributor to the process. This project seeks to ensure his involvement so that his needs are clarified.

3. To improve understanding of the decision process in design so that appropriate design aids, including computer-based aids, to improve design efficiency can be developed.

4. To establish the effects on process time and cost of building regulations with the initial emphasis being on the implications of exempting smaller works from control. This is a subject which has been identified as important during this Forum. London apart, the United Kingdom has a national set of building regulations applied by local building officials. I was a little surprised that the possibility of certification by professionals covered by insurance was not mentioned as an alternative to detailed inspection and checking by independent officials. Our particular concern is to establish both the savings and the additional risks introduced by changes in the regulations.

5. To develop improved systems for communicating project information, particularly to ensure that the designer's intentions are clearly presented to the contractor and his operatives.

6. To develop improved methods for organizing building projects. Planning and scheduling have been mentioned as offering an important opportunity for improving productivity, and our studies are tending to concentrate in the less predictable rehabilitation area, which, as I have mentioned, is a major feature of our current demand.

7. To reduce the level of material waste on site by better design and better organization.

8. To increase productivity by feedback that gives more realistic measures of input and output and creates a better understanding of the relationship between design and production.

Most of these projects are concerned with reductions in time or cost, the denominator of the productivity equation, but we also are concerned with the numerator, the quality and value for money of the finished building. Some of the projects in this latter area have the following objectives:

1. To discover which aspects of design affect the cost of activities housed and the advantages of flexibility and adaptability. The importance of the operational costs of a building also has been mentioned during this Forum as have been total life-cycle costs. This project deals in both of these areas.

2. To improve knowledge of where and why measurable aspects of quality fall short of expected standards and, at the simplest level, to explore defects or below-standard performance which occurs as a result of both design and construction faults.

3. To assess the contributions of project information and site supervision and inspection to the maintenance of quality standards. The advantages of the pre-job conference already have been mentioned. One of the major outputs of our early work on site supervision and its relation to quality is that the site with good communication between trade foremen, general foremen, the operatives themselves, the designer, and his representative on the site is a site where you are likely to get good quality and the quick solution to problems as they

arise. But where people work in watertight compartments and attempt to solve problems themselves, they are not so successful in this regard.

This is a sample of the problems that we identify as important and are studying. Our aim is to develop not only better understanding of a complex process but also quite specific advice, guidance, and standards for application.

Now for a specific suggestion, and it is here that I enter dangerous ground. A number of comparative studies have shown that site productivity in the United States is much higher than in the United Kingdom but that there is considerable variability. I do not think that my suggestion is absolutely specific to the United Kingdom. I suggest that there is a need to improve feedback both of a qualitative and quantitative nature at all stages of the process but mainly between design and production.

We have carried out a large number of studies on site. All too often the design is such that production on site cannot be efficient because the designer has either not understood or not cared about the influence of the design on production. Often this is demonstrated by the fact that individual trades must make repeated visits to each part of the building to complete their tasks. The resulting interference between them leads to more waiting time, more nonproductive time. Designs can be rationalized to achieve greater productivity--a 10 percent reduction in labor expended is well within reach without using inappropriate solutions or affecting the performance of the building. In other studies, defects are designed in or details are difficult to construct and are, thus, both time consuming and liable to fault. We believe that a major increase in quality and value and a substantial decrease in production cost could result if designers applied a better understanding of the process of constructing buildings. Clearly a great deal of construction work is done in projects where an ideal development sequence is not possible, but in home building, for example, there are opportunities for this approach to be adopted.

Feedback also is important as part of a development sequence to allow early design deficiencies to be corrected, perhaps by identifying operations requiring a high labor input which might be reduced by better design.

Last, feedback is important for planning and control of site operations, particularly to establish operational times and to identify levels of nonproductive time and associated reasons and, thus, to allow more effective resource deployment.

The quantitative element in this feedback has not proved easy to obtain in building. I believe it is fair to say that this point has been discussed a number of times during this Forum. We have developed the activity sampling approach and now have a computer-oriented general purpose tool for measuring labor expenditure on site. The wide application of this approach would, in my view, be of great benefit.

The method depends upon making regular (perhaps hourly) snapshot observations of what each operative on site is doing (by selection from a hierarchically ordered set of operations and activities) and his location (by block, house, etc.). The proportion of observations of a particular operation will equal the proportion of time on that operation, although, of course, the accuracy will depend upon size of sample and, hence, the observation frequency. These observations are recorded on a standard form that provides computer input. Figure 1 shows the type of form and indicates the type of information collected. A documented suite of programs is available for the production of a range of tabulations and graphs. The computer output can be in actual man-hours.

From such analyses can be determined detailed breakdown of man-hours to activities, and Figure 2 shows the results for a particular and not untypical site. The presentation emphasizes two major subdivisions of activity time--on the left, the activities that make a positive contribution to the growth of the building and, on the right, those that do not. Among the latter are nonproductive time, some of which must be inevitable, and supervision time, which would be accepted as an essential. At a more detailed level we can compare the labor requirements for sequential design solutions for a particular operation. In Figure 3, the man-hour requirements for the individual blocks of houses in a scheme for the installation of services are shown on a time base. Three visits by the trade generally are necessary and the total man-hour requirement was about 38. Figure 4, shows that a redesign towards rationalization reduced the number of visits to 1 and the number of man-hours to 22. Incidentally, this figure is one of the forms of computer output available in our package. Figure 5 is an example of a redesign indicated to be desirable by feedback. This prefabricated service entry component allows for entry of gas, electricity, water, and telephone without interference with other operations--not only is time saved but the critical path is made much more straightforward.

Figure 6 illustrates the progress over a particular site. The early stages associated with the basic shell, having been rationalized, proceed on a smooth flow but the later stages, designed without thought of production efficiency, involve multiple visits by tradesmen and make smooth progress impossible.

I am therefore suggesting that feedback is essential but that it must be factually based on observation and analysis. I remember it being stated earlier that too often a purely qualitative visual appreciation of productivity was made and used as the basis for decisions. Perhaps wider application will be more successful if it involves all members of the building team. Indeed, I would echo the general theme that communication between members of the team is vital. In the building process, no man is an island. It is relevant to various stages in the design-production part of the process. Of vital importance, at least in the United Kingdom, is the feedback from production to design--as part of a development sequence if possible and as part of architectural education in terms of general principles certainly. We now are taking initiatives to demonstrate how this might be done. Our studies convince us that great benefits can stem

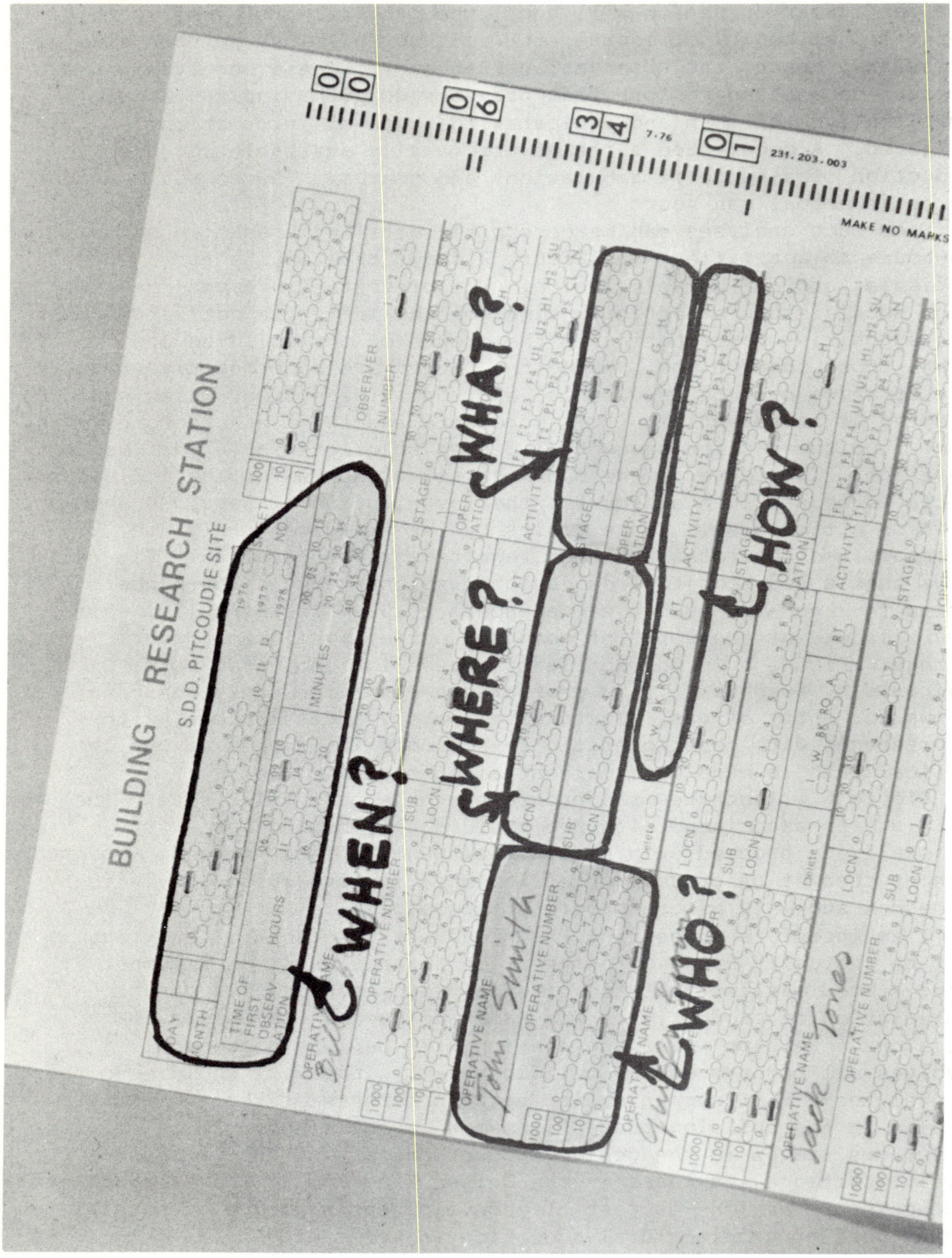
BUILDING RESEARCH STATION
S.D.D. PITCOUDIE SITE
WHEN?
WHERE?
WHAT?
HOW?
WHO?
MAKE NO MARKS
231.203.003
7-76
OPERATIVE NAME
OPERATIVE NUMBER
HOURS
MINUTES
OBSERVER NUMBER
TIME OF FIRST OBSERVATION
STAGE
ACTIVITY
LOCN
SUB LOCN

FIGURE 1

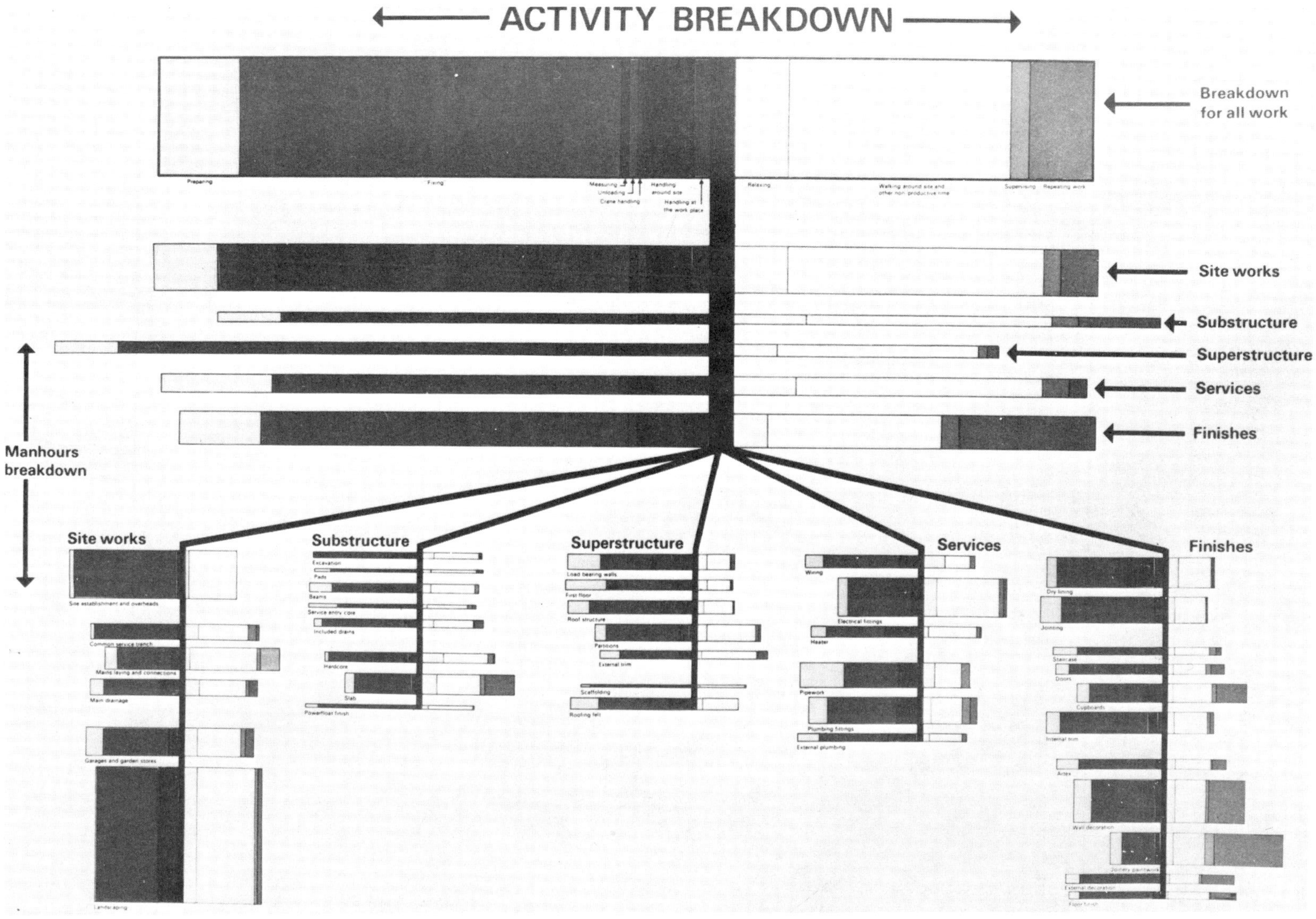
ACTIVITY BREAKDOWN
Breakdown for all work
Preparing
Fixing
Measuring
Unloading
Crane handling
Handling around site
Handling at the work place
Relaxing
Walking around site and other non productive time
Supervising
Repeating work
Site works
Substructure
Superstructure
Services
Finishes
Manhours breakdown
Site works
Site establishment and overheads
Common service trench
Mains laying and connections
Main drainage
Garages and garden stores
Landscaping
Substructure
Excavation
Pads
Beams
Service entry core
Included drains
Hardcore
Slab
Powerfloat finish
Superstructure
Load bearing walls
First floor
Roof structure
Partitions
External trim
Scaffolding
Roofing felt
Services
Wiring
Electrical fittings
Heater
Pipework
Plumbing fittings
External plumbing
Finishes
Dry lining
Jointing
Staircase
Doors
Cupboards
Internal trim
Artex
Wall decoration
Joinery paintwork
External decoration
Floor finish

BRE SITE ACTIVITY ANALYSIS PACKAGE SITE ID 4 DATE 08/10/79 TIME 17/49/08 PAGE 4

PAGE PARTITION 1.1

PAGE SELECTION NUMBER 6 OPERATION CODE 025A WIRING CONDUIT BOXES
ACTUAL RANGE OF VALUES FOUND: 025K -025A

ROW SELECTION NUMBER 1 DATE
COLUMN SELECTION NUMBER 3 BLOCK

CODE	BLOCK 0	B1	B2	B3	B4	B5	B6	B7	B8	B9	B10	TOTAL
						CLASS CODES						
6 375												0
10 375												0
17 375												0
24 375												0
31 375												0
7 475												0
14 475												0
21 475												0
28 475												0
5 575												0
12 575			27		1							29
19 575			36									36
26 575		14	42									56
2 675		48	10	5								63
9 675		11	39	16								65
16 675		28	16									44
23 675		16	19	16	3							54
30 675			17	39	15							71
7 775					34	21						55
14 775					35	37						72
21 775						3	64	1				69
28 775												0
4 875												0
11 875						26						26
18 875						9						9
25 875			1			22	18					41
1 975		1						5				6
8 975				1		2	48	4				55
15 975			1			20	21	64				106
22 975							1	74		6		82
29 975		9	44					33				86
61075		2	12						19			32
131075		31	28									59
201075		18	22	13					8			62
271075		7						10	55			72
31175		27	8	22					1			58
101175		7	1	57						3		67
171175				52						3		55
241175				3	1	2	1			48		54
11275			1		26	5				25	5	62
81275					25	1	1			2	23	51
151275					6	21					17	43
221275						13						13
291275												0
5 176						9					42	51
12 176					11	35					1	47
19 176					2	17	25	1			1	46
26 176					1		19	13				33
2 276			3		13	7	14	2				40
9 276							37	14				51
16 276		1	14				6	18			5	45
23 276			15				3	26				44
1 376		7	13	1			16	9	14			54
8 376		6		33	6			1	3	8		55
15 376		2			14	8	1			26		51
22 376		4	3	11	6	27		8				58
29 376		1	3	5	7	18	3	6				42
5 476						1	25	32	9			66
12 476							20	29	3			53
19 476								41	11			52
26 476								56	14			70
3 576								9	35	8		52
10 576							1	2	1	18	38	60
17 576										44	41	85
24 576											1	1
+												
TOTAL	0	238	373	272	206	304	325	459	172	189	174	2709

FIGURE 3

BRE SITE ACTIVITY ANALYSIS PACKAGE SITE ID 77 DATE 11/10/79 TIME 17/55/14 PAGE 4

PAGE PARTITION 1.1

PAGE SELECTION NUMBER 6 OPERATION CODE 019A WIRING AND OUTLET BOXES

ACTUAL RANGE OF VALUES FOUND: 019A -019A

ROW SELECTION NUMBER 1 DATE

COLUMN SELECTION NUMBER 3 BLOCK

:BLOCK / CODE	0	BL1	BL2	BL3	BL4	BL5	BL6	BL7	BL8	BL9	BL10	BL11	BL12	BL13	BL14	BL15	BL16	BL17	BL18	BL19	BL20	BL21	BL22	BL23	BL24	BL25	BL26	BL27	BL28	BL29	TOTAL
18 876																															0
23 876																															0
30 876																															0
6 976																															0
13 976																															0
20 976																															0
27 976																															0
41076																															0
111076																															0
181076																															0
251076																															0
11176																															0
81176																															0
151176																															0
221176																															0
291176																															0
61276																															0
131276																														15	15
201276																												1	5	6	12
5 177																														6	6
10 177																															0
17 177																													1	90	91
24 177																												1	13	82	96
31 177																													40	46	86
7 277																												1	95		96
14 277																												54	25	1	80
21 277																										2		77	1	3	83
28 277																					1	4	2	84	69	8		2		169
7 377																1						35		59		41				136
14 377																3	1	3		23	1	80	44						1	155
21 377																		54		42	46		23							165
28 377																1		62	34	1	30									128
5 477																1	1	88	17			1								107
11 477																120	18	6	7		1	1		1			1		1	155
18 477													19	25	20	73	8	1					1	1						152
25 477								1	14		98		24	25	4				1					1		1				169
2 577						19	25	35	13	43	3	16				1		2				1				1				160
9 577					11	57	1	18		18	2	37		3		3		1				4	6	1						161
16 577		1	30	27	27	3							26	1		2		2												118
24 577		74	2						1				2																	79
30 577		40			1																									41
6 677		5	1			2	1		2	1		2	1	1		1														17
13 677		3		1		3	1	1	5	1	3																			19
20 677											2				1															3
27 677																															0
4 777																															0
11 777																															0
1 877																															0
8 877		1	1				2		4																						8
15 877					1																										1
22 877																															0
TOTAL	0	124	34	28	29	19	80	27	66	30	66	105	84	45	53	25	209	28	219	59	66	79	125	82	147	71	51	135	181	251	2511

FIGURE 4

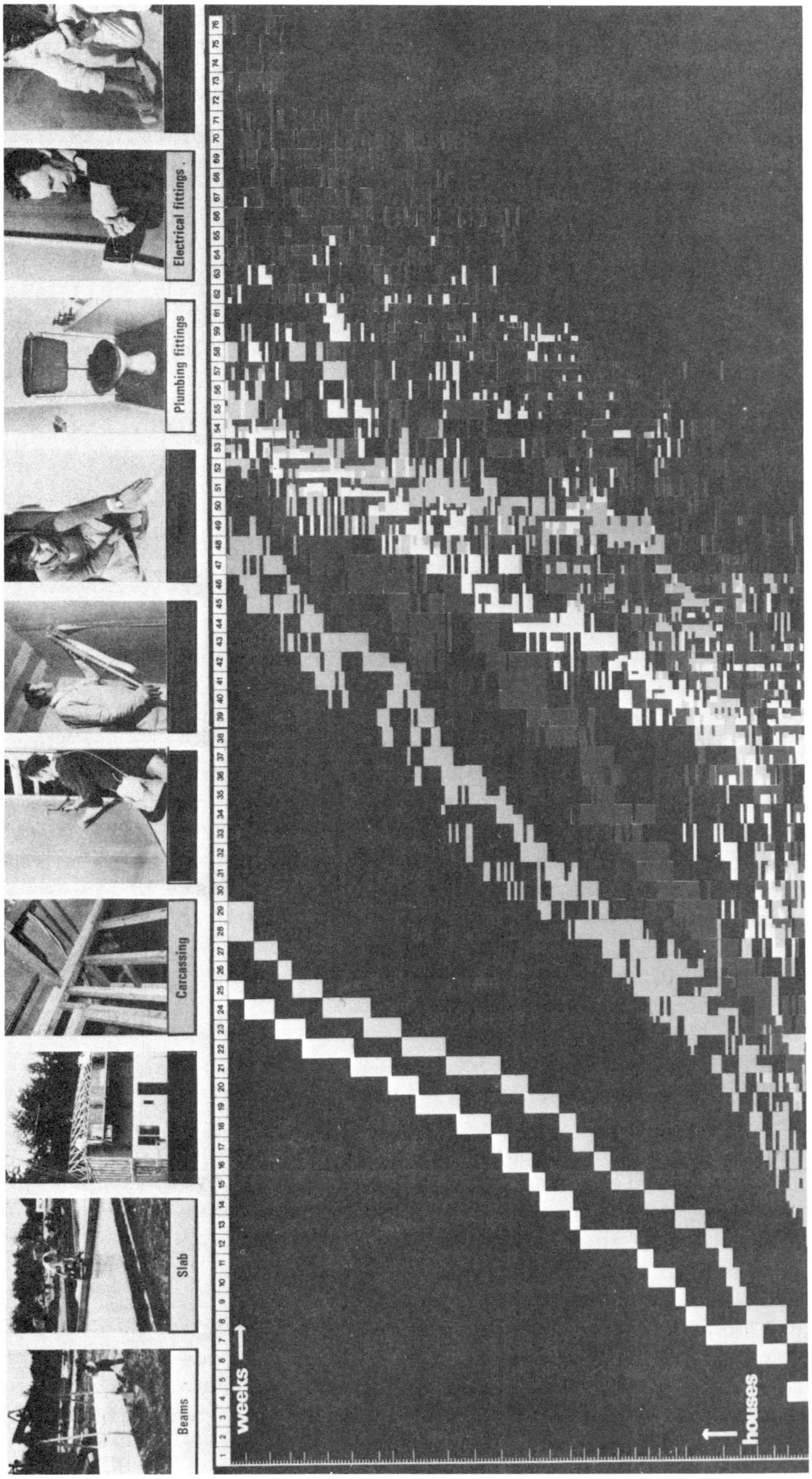
Beams
Slab
Carcassing
Plumbing fittings
Electrical fittings .
weeks
houses

FIGURE 6

from designs which are straightforward to construct, easy to organize, and avoid in-built faults; that sequential development pays off; and that realistic site planning is worthwhile. With respect to all these requirements, there is a need for a realistic measurement of labor expenditure on site so that development can be founded on fact. We believe we have the essential features of such a management tool now available for wider application.

JOJI ARAI
Manager
Japan Productivity Center, Washington, D.C.

Almost every day you read in most of the magazines and newspapers something about productivity. It was not that way a few years ago. All of a sudden productivity has become an important issue for everybody.

When some talk about productivity, they often give statistics to show that Japan's productivity grew at the rate of 8.9 percent in the past two decades while U.S. productivity grew only 3.8 percent in 1950 and 2.3 percent in the 1960s. The U.S. figure now stands at about 1.4 percent. Such people try to alarm you, but I am sure others have told you that the per capita gross domestic product of the United States stands at 100 percent, which is the highest in the world, while Japan's stands at only 65 percent. Thus, Japan still needs to catch up and that is the reason why we are growing faster.

Many factors contribute to the rapid growth of productivity in Japanese industries, and these include:

1. Government policies and programs that actively and passively support the economic expansion,
2. Capital investment,
3. Technological innovation,
4. The corporate management tendency to take long-range views in policy formulation,
5. The rising educational level of the work force,
6. Improved skills and work ethics of employees,
7. Heavy emphasis on quality control and assurance,
8. Harmonious labor-management relations,
9. Economy of scale, and
10. Systems improvement.

With respect to government policies and programs, Japanese industry benefits from special tax measures, accelerated depreciation, and other preferential policies and programs toward growth. The government tends to offer assistance to high potential, high growth, high productivity industry and to disregard low productivity industry, which consumes large amounts of energy and material, because Japan has no natural resources other than people. Japan places heavy emphasis on the human aspect of productivity. Because of the substantial accumulation of wealth and abundant supply of natural resources in the United States, you place heavy emphasis on investment, hard technology, and research and development efforts.

We, however, also recognize the importance of capital investment. In 1978, for example, Japan invested $144 billion in new plants and equipment and the United States invested $148 billion. The gross national product of Japan is one-half that of the United States, and the size of Japan is one-twenty-fifth of the United States. Thus, you can imagine the concentrated effort Japanese industry is making in the area of investment in equipment.

Similarly, Japan recognizes the importance of technological innovation. Last year, we spent $20 billion on research and development programs. The United States, on the other hand, spent $40 billion. Both countries spent about 2 percent of their gross national product on research and development programs, and on this score, we are on a par.

Japan has what we call a "lifetime employment system." In most corporations, once a man gets a job, he tends to stay with the company until he reaches the age of retirement at 60. Ups and downs of the economy and temporary slacks in performance do not necessarily result in the discharge of a man from the corporation. This policy and practice applies to mangement as well as blue collar workers; therefore, senior management, particularly the chief executive officer, can make decisions based on a long-term viewpoint. The goal for a Japanese organization is to expand in the future, not to seek immediate return on investment or short-term profits. In this way, I think there is a substantial difference between American and Japanese management.

As for education, you know that level of education has a great deal to do with the relative ease or difficulty with which technological innovations can be implemented. The Chinese are learning this at the present time. They bought a new steel plant from us but did not know how to run it. Now, they are recognizing the importance of training managers, but it will be years before they can get performance comparable to ours out of the facility.

You often hear about the quality standard of Japanese performance. Let me remind you that the Japanese learned the importance of quality control from Americans. In the 1950s, Drs. Duran and Deming first exposed Japanese management to the statistical and total quality control concept. Dr. Duran's concept was to involve middle managers in the quality control program. The Japanese expanded the concept and involved the blue collar workers. This new system is called the "Q C Circle Program." It is a very simple thing. The workers are taught how to draw cause and effect diagrams and histograms and sometimes even to do regression analyses, something taught in the United States to graduate-level students. These things are taught to the high school graduate at the working station and he then knows exactly how to analyze his operation. He finds out exactly what is wrong. If any defective product comes out of his working station, he analyzes the cause and comes up with the solution. Basically, this program is only a brainstorming session but its use has resulted in Japan's obtaining the reputation as a producer of high-quality goods.

One example might interest you. A Japanese electronics company purchased a division of a very large American electronics company in the Midwest. Under American management, 15 to 18 out of every 100 color television sets packed usually were found to be defective by the customer. After the Japanese took over, the defective ratio was reduced to 4 per 100 units. This is still very high, however, since in Japan the ratio is 0.4 percent. That illustrates the importance we place on quality.

The harmonious labor-management relations that result from our lifetime employment system virtually guaratee management the long and loyal service of employees. On the other hand, the employees have the guarantee of jobs and the advantages of paternalistic programs, and they share the company's destiny on a long-term basis. Therefore, Japanese workers are motivated because they know their future is almost identical to that of their company.

Americans also taught us the importance of applying the principle of economy of scale. I therefore do not think I have to go into the details.

I would like to mention one thing about systems analysis and systems improvement. Japan now is placing tremendous emphasis on the production of robotic systems. In fact, about 40,000 robots now are operating throughout Japan in various industries. One company has developed an automatic sweeper (I do not know if there is a comparable one in the United States) that has a great deal to do with building maintenance costs, which I understand represent 6 percent of building costs. This automatic sweeper is a robot. It has its own sensing unit mechanism that automatically switches on and off. When it hits an object, it changes course. Turn one of them loose and it automatically cleans the floor space. The robot behaves exactly as programmed when you have only one in a room, but if you put several in the same room, they behave quite differently from the way they are programmed. They somehow select a leader, and then the rest follow it. It also has been found that some robots made from the same parts and using the same manufacturing process have more sensitive perceptions than the others--i.e., their sensing units somehow excel over the others--and some robots function a little bit better mechanically than the others. It also has been found that the mechanically superior robots tend to be leaders whereas those with sensitive sensing units tend to be followers. Maybe robots are trying to tell us something.

In closing, I would like to quote the declaration made by the European Association of National Productivity Centers in 1950 that pertains to the interpretation of productivity: "Productivity is a state of mind. Mind being confident that tomorrow can be made better than today." We tend to interpret productivity as output divided by input, but I wanted to give you this interpretation of productivity as well.

DENNIS J. SULLIVAN
Vice President
American Productivity Center Inc., Houston, Texas

The American Productivity Center is a privately funded, nonprofit organization that is vitally concerned with productivity in all facets of the American economy. We conduct a series of educational and informational programs to bring to the attention of various industry and public sector groups the need for improvements in our national productivity and in individual organizations' productivity. That concludes my corporate "plug."

During this Forum we have heard quite a bit about productivity and the need for productivity improvement in the construction industry. It would certainly seem that anything important enough to have caused the Building Research Advisory Board (BRAB) to convene a meeting like this and for people like yourselves to attend it would be understood and that all parties--management, labor, and government--would be highly motivated to increase productivity. Yet, the sad fact is that this is just not true. Many facets of labor oppose productivity improvement because they see it in terms of a "speed up" and a return to the less enlightened approach to labor-management relations of the past. Or they attribute it to simply a measurement anomaly; many say we have a measurement not a productivity problem.

I do not agree and my organization does not agree. Our measurement tools may not be as precise as they need to be and, within an industry like construction, there may be room for quite a bit of research on measurement, but, in fact, we do have a productivity problem.

On the government side, although there are some hopeful signs, we still have to contend, and you have to contend very directly, with the burdens imposed by two decades of productivity-strangling legislation and regulation. Dr. Weidenbaum's panel addressed itself to that subject and gave you a feeling for the enormity and the scope of the problem.

As unfortunate as the government problem is, it is nonetheless far easier to rationalize than the lack of understanding and constructive productivity improvement programs in management. To be sure, every one of you is in favor of productivity improvement or you would not be here. Very little in your own organizations, however, is being done about productivity improvement.

Let me ask you several questions. How do you define productivity? What does it mean in your own organization? What is the productivity level of your own organization or firm? How do you measure that productivity? What are the productivity inputs in your own organization or your segment of the industry? Does your organization have a stated set of productivity goals? Do you have a productivity program internal to your organization? How do your employees feel about that productivity program?

I would be very, very surprised if more than 20 percent of the audience could answer those questions. The problem of productivity and its relation to the economic health of our nation and of your industry is recognized. We do need to conduct research into the cause and effects of declining productivity in this country. My organization is one of those that is involved in that.

More importantly, I believe it is paramount that your own organizations, your individual organizations, undertake positive programs to improve productivity. You cannot and should not look to government or even something like BRAB for the answers to your own productivity problems. The answers can be and should be developed within your own organizations. Regulatory reform and legislative actions are needed to create some improved climate for productivity improvement, but the key to it is internal to your own organization.

Let me discuss the factors that we believe are essential to the establishment of successful productivity improvement programs. Top management involvement is of great importance. A successful program cannot be started anyplace else but in the front office, and the involvement there must be more than token; it must be continuous over the life of the program if it is going to be a success.

Organization also is essential. It must be inherent in the structure of your organization. If that means modifying your organization in order to increase productivity, you had better do it.

Productivity has a multiplicity of definitions, and I am particularly fond of the one with which Mr. Arai closed. However, we feel it is very important that the individual organization define productivity in quantitative terms to the extent possible so that we can get into the next facet of a proper productivity program and that is measurement.

You cannot measure what you have not defined. It is entirely possible and probably quite likely that you will not have a single measure of productivity or a single definition of productivity for anything as complex as a construction organization. There is nothing wrong with having multiple measures. It is important to measure, however. You need to know where you are before you begin any sort of an improvement program and you need to be able to assess what kind of progress you are making.

It also is important that productivity and productivity improvement become an integral part of training. I do not care whether it is basic skills training at the apprentice level or supervisor and management training. The importance, the criticality, of productivity to the health and growth of the organization needs to be emphasized at all levels of training.

Productivity growth goals also must be established and periodically revised so that the organization knows whether it has achieved positive or negative growth and so that it will continue to have something to strive for. Dr. Jeanes mentioned the importance of feedback. Inherent in the establishment of feedback within an industry, within an organization, is the existence of some goals, some benchmarks, against which to report feedback.

The fact that you have a program, that you are concerned with productivity, needs to be communicated. It needs to be communicated to your individual employees, to the unions with which you work, to every one of your managers. Productivity is not an item to be discussed in a company newsletter only once a month. The program, its scope and its importance, needs to be emphasized and communicated on a regular basis.

Mr. Arai made another point when he spoke about quality control in Japanese industry. The key to it, in his eyes, is employee participation. We believe very strongly that productivity programs cannot exist, cannot be successful, without direct employee participation. The employees are probably, as Mr. Georgine stated earlier, the experts on the individual elements that make up the job. They know more about each discreet element on a construction job than we do. Again, in deference to Mr. Arai, people are the important resource. They are the only resource available in Japan, but they also are a very critical resource in your own organization. This resource needs to be tapped and tapped effectively.

Any program that you undertake for productivity improvement must be evaluated. Since you are going to expend resources on it, you must know if you are expending them correctly, if they are having the kind of impact you want them to have, and if they are having the impact on your productivity that you expected. If they are not, you must know why.

The last point I will make relates to information and information resource sharing. One of the important things that we find ourselves doing for industry right now is acting as a clearinghouse. As Mr. Arai pointed out, there has been a tremendous upsurge of interest in the past year or so in the topic of productivity and how one improves productivity. The question that usually follows is: What can I do and how can I do it? I think one of the important things that will or should come out of this meeting is the fact that information on effective productivity programs, on productivity-enhancing technological innovations, needs to be disseminated in the industry. That implies that it is being collected somewhere and that it is retrievable. I think that might be a very legitimate activity for this Forum to recommend as an output.

The Joint Economic Committee of Congress in a report issued in August 1979 made the following statements:

> The average American is likely to see his standard of living drastically decline in the 1980s unless the United States accelerates its rate of productivity

> growth.... It is emphasized that if no new steps are taken to address the problem of structural unemployment, lagging capital formation and the slowdown in productivity then the American economy faces a bleak future. A stagnating economy will mean fewer Americans will be able to afford the necessities of life, such as a decent home.

That strikes at the heart of your industry and it graphically expresses why you should be working on productivity in your own organizations at this very moment.

L. KENNETH HARMON
Corporate Productivity Coordinator
Mead Corportation, Dayton, Ohio

I will try to provide some insight into the productivity improvement process as we have been experiencing it at the Mead Corporation. I should note, however, that my observations and comments no doubt will be influenced by my having spent a year on the staff at the American Productivity Center.

I would like to make four suggestions for the building and construction industry on the subject of developing a productivity improvement strategy. First, concentrate on the management of the individual firm and, within the firm, suggest that managers concentrate on individual profit centers. Do not dwell on the national productivity problem or even the low productivity of the building and construction industry. There is nothing wrong with using a little patriotism or even a little industry pride to get people looking in the right direction. However, there is very little corrective action that any of us can take at those two levels of aggregation. Productivity improvements actually occur in the profit center operations of the individual firm. Government agencies and industry associations can observe and study the problem and even can establish goals and overall policy, but the most useful thing they can do is to encourage the managers of individual firms to launch productivity improvement initiatives that focus on profit center operations and then to be supportive of these managers in their local efforts to achieve larger, more frequent productivity gains.

This approach also works for the top management of large corporations. The more successful ones do not even try to directly control productivity from the top; that is what their profit center managers were hired to do. However, if improvement is the goal, top management must take a leadership position on the productivity issue and encourage their profit center managers to establish local productivity improvement initiatives. As in the case of government and industry associations, the top management in corporations also must follow through by supporting operating managers in their local efforts to improve productivity.

My second suggestion is that everyone involved must be helped to recognize that we are talking about an improvement problem. By definition that means we are looking for more and better than we have experienced in the past. In productivity improvement, we are talking about an "over and above" effort to achieve goals that "stretch" our present capabilities. More business as usual, without significant changes, simply will not solve the improvement problem. Working smarter on the job site and in the office is what is needed; only infrequently is working harder required. Also important is getting more out of the time spent on the job and the equipment purchased for the job. Needed is an attitude of always improving on yesterday's performance.

In any industry, including building and construction, an understanding and a commitment to the philosophy of improvement must underlie all initiatives to solve the problem of declining productivity. Again, I would suggest that government, industry associations, and corporate executives have an important part to play although managers and employees at the profit center level are the people who ultimately will solve the problem. The top level leadership, in this case, must explain and gradually impart an understanding of the improvement nature of the solution to the productivity problems. It is necessary to develop commitment to decisive action at the profit center level, not to engage in further rhetoric in high places.

Productivity improvement requires an extra effort, but an extra effort on the part of individuals who are already fully occupied and frequently very busy must be well organized and well managed to produce the desired results without disrupting the normal flow of basic operations. Most firms and profit centers need a way to organize and control this kind of "over and above" effort. Many firms are finding that a productivity improvement program is an effective management technique for achieving these objectives.

My third suggestion is that everyone involved must believe in and feel comfortable with the fact that a good profit center manager will be committed to the extra effort required only if he can first see a profit in his actions. This attitude in a free enterprise economy is good, proper, and commendable. Profit is good! The principal reason the general public is concerned about corporate profits is because they have come to associate large profits with inflationary prices. Profit itself is not bad. Productivity improvement is a noninflationary source of greater business profits.

Productivity gains produce rewards for everyone. The profit center manager gets his profit increase. The public does not have to be burdened with further price increases. Business owners are able to finance growth which produces more jobs and so on. If the people involved believe there are big rewards associated with productivity gains and if they understand that change is required to create the needed improvements, I believe that anything further we can do to help the profit center manager will be a positive input to the ultimate solution of a very complex problem.

Our recent experience with productivity improvement leads me to make a fourth suggestion. The leaders in any productivity improvement effort should concern themselves with the long-term development of productivity management capability. They should discipline themselves to concentrate on the strategic and long-term aspect of solving the problem, not the short-term firefighting aspect. Every industry, every firm, every profit center organization already has several very qualified firefighters and ramrods. They are called shift foremen, supervisors, and crew chiefs in some organizations.

Productivity has declined in recent years because executives and managers have not been working very hard to make it climb. Our productivity management in this country is rather weak and ineffective. On the other hand, to strengthen our productivity management capability will not only solve the immediate problem of productivity declines but also will ensure that the problem will remain solved for a long time. We must all remember that employees and assets make great things possible in business, but management makes them happen. It is effective management that breathes life into lazy assets and motivates average workers to be change agents. One of the more useful management tools that can be employed by profit center managers in their efforts to generate greater productivity gains is, once again, a productivity improvement program.

I have made four suggestions for an approach to improving productivity in the building and construction industry. You might say I have recommended a FIRM strategy, the letters in the word FIRM helping me to remember these four suggestions:

F - Concentrate on the management of individual <u>Firms</u> and within the firm suggest that managers concentrate on profit center operations. Focus on the individual building and construction firm.

I - Help everyone concerned understand that we are involved in an <u>improvement</u> situation which demands that we make changes on a continuous basis. Explain the improvement nature of this situation to all employees in the industry. Create individual commitment to participate.

R - Help everyone involved understand that productivity gains mean big <u>rewards</u> for us all. Productivity gains are a noninflationary way to increase profits. Explain the rewards of productivity improvement to all employees in the industry. Create employee support for management initiative.

M - Concentrate on the development of productivity management capability. Do not try to manage the other guy's organization. Just help him become more effective in his management of future productivity gains.

At this point I would like to share with you what we believe to be the underlying process of a successful productivity improvement program:

1. Executive Commitment--Desire for organized corporate approach and decision to implement.
2. Program Management--Accountability for directing and supporting the special effort to create further productivity gains.
3. Accurate Definition--"Total" productivity linked to profitability.
4. Measurement Consensus--How they will determine the success or failure of productivity improvement initiatives.
5. Improvement Goals--Specific "stretch" goals for each operating manager.
6. Awareness Training--Inform and motivate employees of the need for individual effort and overall productivity gains.
7. Creative Supervision--Productivity-oriented leadership with a sensitivity for the QWL needs of productivity-oriented employees.
8. Objective Appraisal--Identification of problems and opportunities for further productivity gains.
9. Participative Planning--Involving employees in the process of planning how future operations will deal with the major problems and opportunities.
10. Adequate Resources--Staffing and equipping the organization for further productivity gains.
11. Performance Training--Productivity-oriented training to provide job skills for employees at all levels.
12. Incentive--Providing all forms of compensation and reward for initiating and participating in the changes required to increase productivity.
13. Performance Evaluation--Evaluation of organizational performance in terms of specific productivity improvement objectives, plans and standards.

I am sure you have recognized that this is really just an organized approach to managing improvement of any type. The exact number of steps in the process is unimportant. What is important is that profit center managers be encouraged to establish an organized local approach. They may choose to call it a productivity improvement program or they might prefer a less formal title.

Probably what is needed more than anything else is to simply encourage the managers of individual building and construction firms to establish productivity improvement programs and then to support these managers in their efforts to do so. The support can come from corporate executives, industry associations, and government agencies, but the local profit center manager must establish and direct any organized effort to increase productivity. I believe the process described above is a description of the way most managers would go about the task, given the time and research input to develop a program from scratch. Rather than that, why not start with our process and evolve your own refinements as you move forward? There is really no time to spare in reinventing the wheel.

The American Productivity Center also has developed resources and a professional staff that is highly qualified to assist managers in establishing local productivity improvement programs. Why not tap their experience in your efforts to encourage individual building and construction firms and in your efforts to support these managers with specific, useful input?

APPENDIX A

LIST OF FORUM PARTICIPANTS

Jack E. Abbott, Manager Staff Services, PPG Industries, Pittsburgh, Pennsylvania

Gifford H. Albright, Professor and Academic Department Head, Department of Architectural Engineering, Pennsylvania State University, University Park

Louis Alfeld, President, Decision Dynamics Inc., Reston, Virginia

John J. Andrews, Vice President, Smith Hichman and Grylls, Detroit, Michigan

Joji Arai, Manager, Japan Productivity Center, Washington, D.C.

William Bahrke, President, Dravo Utility Constructors, Inc., New York, New York

David J. Baker, Executive Vice President, Cardinal Industries, Columbus, Ohio

Robert E. Berger, General Physical Scientist, National Bureau of Standards, Washington, D.C.

William S. Birney, Marketing Manager, U.S. Steel Corporation, Washington, D.C.

M.S. Blackistone, Consultant, Building Research Advisory Board, Washington, D.C.

Raymond C. Bliss, Research Assistant-Statistician, International Union of Bricklayers and Allied Craftsmen, Washington, D.C.

Sanford E. Blumenthal, Manager, Managering Services, National Electrical Contractors Association, Inc., Washington, D.C.

Seymour A. Bortz, Senior Engineering Advisor, Materials Technology Division, Construction Productivity Center, Illinois Institute of Technology, Chicago

James Brown, Professor of Law, National Law Center, The George Washington University, Washington, D.C.

Cynthia Burton, Program Associate, Maryland Center for Productivity and Quality of Working Life, University of Maryland, College Park

William F. Caldwell, President, Caldwell Equity Corporation, Troy, Michigan

Robert Chapman, Operations Research Analyst, National Bureau of Standards, Washington, D.C.

Pamela J. Clayton, Vice President, BOSTI, Buffalo, New York

Robert P. Connolly, Manager, Technical Services--Government Relations, Washington, D.C.

Charles G. Culver, Senior Staff Engineer, National Engineering Laboratory, U.S. Department of Commerce, National Bureau of Standards, Washington, D.C.
William J. Curtin, Senior Partner, Morgan, Lewis and Bockius, Washington, D.C.
Charles J. Dinezio, Executive Director, State Building Code Commission, Boston, Massachusetts
G. Day Ding, Head, Department of Architecture, University of Illinois, Urbana
Porter Driscoll, Manager, Technology Applications, National Bureau of Standards, Washington, D.C.
Robert Martin Engelbrecht, Robert Martin Engelbrecht and Associates, Princeton, New Jersey
Joan D. Finch, Executive Secretary, Technology Assessement and Utilization Program, Building Research Advisory Board, Washington, D.C.
Thomas J. Flynn, Jr., Senior Vice President, Real Estate Loan Department, Bank of America, San Francisco, California
Maurice H. Gardner, Manager-Projects Operation, General Electric Company, Schenectady, New York
Martin D. Gehner, Professor of Architectural Engineering, Yale University, New Haven, Connecticut
Robert A. Georgine, President, Building and Construction Trades Department, AFL/CIO, Washington, D.C.
Richard P. Godwin, Executive Vice President, Bechtel Incorporated, San Francisco, California
Thomas M. Gose, Office of the Secretary, U.S. Department of Commerce, Washington, D.C.
Paul C. Greiner, Vice President, Conservation and Energy Management, Edison Electric Institute, Washington, D.C.
James G. Gross, Chief, Building Economics and Regulatory Technology Division, National Bureau of Standards, Washington, D.C.
John W. Guinee, Jr., Real Estate Investor, Reston, Virginia
G. Ralph Guthrie, President, Urban Investment and Development Company, Chicago, Illinois
Samuel L. Hack, Management Consultant, Washington, D.C.
Jeffrey J. Hallett, Director, Productivity Center, Economic Division, Chamber of Commerce of the United States, Washington, D.C.
Irvin Hamburger, U.S. Air Force, Washington, D.C.
Calvin S. Hamilton, Director of Planning, Department of City Planning, Los Angeles, California
Robert C. Holland, President, Committee for Economic Development, Washington, D.C.
John C. Horning, Manager-Engineering, General Electric Company, Schenectady, New York
R. E. Jeanes, Head, Planning and Construction Department, Building Research Establishment, Garston, Watford, England
Oliver H. Jones, Consulting Economist, Oliver Jones and Asociates, Manns Choice, Pennsylvania

F. S. Kellstrom, Chairman of the Board, Fischbach and Moore, Inc., Los Angeles, California

Charles H. Kimzey, Cooperative Technology Program, National Bureau of Standards, Washington, D.C.

C. Lee Kitchin, Vice President of Construction, Davy McKee Corporation, Independence, Ohio

Charles B. Knapp, Deputy Assistant Secretary of Labor for Employment and Training, U.S. Department of Labor, Washington, D.C.

Wilbur F. Koepke, Manager, Real Estate and Property Management, American Telephone and Telegraph Company, Basking Ridge, New Jersey

Marvin Kraettli, Senior Training Coordinator, Brown and Root, Inc., Houston, Texas

Creighton C. Lederer, Director and Commissioner, Buildings and Safety Engineering Department, Detroit, Michigan

John W. Leonard, Vice President, Morrison-Knudsen Company, Inc., Boise, Idaho

Thomas E. Lewis, Manager, Government Marketing and Technical Services, Armstrong Cork Company, Washington, D.C.

Walter H. Lewis, Professor, Department of Architecture, University of Illinois, Urbana

Sam Little, Transportation Coordinator, Environmental Protection Agency, Washington, D.C.

Wesley Long, Deputy Director, Office of Industrial Economics, U.S. Department of Commerce, Washington, D.C.

Thomas J. Lucas, President, TMP Associates, Architects, Bloomfield Hills, Michigan

Lindsay Lucke, Student University of Maryland, College Park

Donald T. Lyon, Manager, Building Planning and Design, American Telephone and Telegraph Company, Basking Ridge, New Jersey

Philip Marcus, Economist, U.S. Department of Commerce, Washington, D.C.

Charles D. Markert, Student, American University, Washington, D.C.

Albert R. Marschall, Commissioner, Public Buildings Service, General Services Administration, Washington, D.C.

Robert Marshall, Jr., Vice Chairman, Turner Construction Company, New York, New York

Paul H. Martinez, Manager, Construction Services, Nestle Enterprises, Inc., White Plains, New York

Frank J. Matzke, Vice President, Office of Technology and Programs, National Institute of Building Sciences, Washington, D.C.

Neil B. McArthur, Vice President, The Austin Company, Washington, D.C.

Gerald T. McIvor, Professor, Illinois Institute of Technology, Chicago

Robert McKinnie, Student Howard University, Washington, D.C.

Gershon Meckler, President, Gershon Meckler Associates, Washington, D.C.

David S. Miller, President, David S. Miller and Associates, Inc., Cleveland, Ohio

Michael F. Mohr, Economist, U.S. Department of Commerce, Washington, D.C.

Dan E. Morgenroth, Director-Market Development, Owens-Corning Fiberglas Corporation, Toledo, Ohio

Get W. Moy, General Engineer, U.S. Department of the Navy, Washington, D.C.

Thelma S. Mrazek, Director of Communications, Mechanical Contractors Association of America, Washington, D.C.

Jon I. Mullarky, Vice President of Promotions, National Ready Mix Concrete Association, Silver Spring, Maryland

Kathleen A. Naughton, Manpower Analyst, U.S. Department of Labor, Washington, D.C.

J. William Nelson, Senior Policy Officer, U.S. Department of Commerce, Washington, D.C.

Scott O'Gorman, Johnson Controls, Washington, D.C.

Richard J. Orend, Senior Staff Scientist, Human Resources Research Organization, Alexandria, Virginia

H. Allyn Parmenter, Training Department, United Association of Journeyman and Apprentices of the Plumbing and Pipe Fitting Industry, Washington, D.C.

C. E. Peck, Executive Vice President, Owens-Corning Fiberglas Corporation, Toledo, Ohio

Roger A. Rensberger, National Bureau of Standards, Washington, D.C.

R. Thayne Robson, Director, Bureau of Economic and Business Research, and Professor of Management and Economics, University of Utah, Salt Lake City

J. L. Rosenberger, Construction Manager, Union Carbide Chemicals and Plastics, Houston, Texas

William D. Rust, Structural Engineer, Public Buildings Service, General Services Administration, Washington, D.C.

Aaron Sabghir, Director, Construction and Building Products Division, U.S. Department of Commerce, Washington, D.C.

Robert F. Schmitt, President, Bob Schmitt Homes, Inc., Strongsville, Ohio

Anita W. Schoomaker, Law Student, Washington, D.C.

L. R. Shaffer, Deputy Director, U.S. Army Construction Engineering Research Laboratory, Champaign, Illinois

Arthur Shinn, U.S. Department of Housing and Urban Development, Berea, Ohio

Irving H. Siegel, Consulting Economist, Bethesda, Maryland

Edward L. Simons, Manager, Environmental Protection Operation, General Electric Company, Schenectady, New York

George L. Smith, Assistant Director, Construction Operations and Facilities Management, Office of the Secretary of Defense, U.S. Department of Defense, Washington, D.C.

Karen H. Smith, Associate Editor, Cahners Publications, Washington, D.C.

Peter H. Smith, Chairman, Gibbs and Hill, Inc., New York, New York

James R. Smith, Senior Staff Officer, Building Research Advisory Board, Washington, D.C.

Ed Stickney, NUS Corporation, Rockville, Maryland

H. Kemble Stokes, Jr., Senior Economist, U.S. Department of Commerce, Washington, D.C.

Dennis J. Sullivan, Vice President, American Productivity Center, Inc., Houston, Texas

John P. Sylva, Assistant Director, Construction Operations and Facilities Management, Office of the Secretary of Defense, U.S. Department of Defense, Washington, D.C.

Hilary W. Szymanowski, Manager-Advanced Conversion Technology, Westinghouse Electric Corporation, Pittsburgh, Pennsylvania

Donald C. Taylor, President, Construction Industry Research, Denver, Colorado

Merlin L. Taylor, Assistant to the President, International Union of Bricklayers and Allied Craftsmen, Washington, D.C.

Bruce Thompson, Economist, National Bureau of Standards, Washington, D.C.

Terry Thompson, Senior Manager, Personnel Services, Brown and Root, Inc, Houston, Texas

Vance W. Torbert, Jr., Architectural Officer, Metropolitan Life Insurance Company, New York, New York

Edwin S. Townsley, Executive Director, Building Research Advisory Board, Washington, D.C.

Warren H. Turner, Engineering Director, American Telephone and Telegraph Company, Basking Ridge, New Jersey

Thomas D. Walczak, Vice President, The E. G. Hauserman Company, Cleveland, Ohio

Martin H. Waldinger, Vice President, The Waldinger Corporation, Des Moines, Iowa

Murray L. Weidenbaum, Director, Center for the Study of American Business, Washington University, St. Louis, Missouri

Arthur M. Weimer, Consulting Economist, U.S. League of Savings Association, Washington, D.C.

Robert M. White, Administrator, National Research Council, Washington, D.C.

Robert C. Wible, Editor and Information Director, National Conference of States on Building Codes and Standards, Inc., McLean, Virginia

John H. Wiggins, Jr., President, J. H. Wiggins Company, Redondo Beach, California

John W. Williamson, Director, American Telephone and Telegraph Company, Basking Ridge, New Jersey

Robert L. Wilson, President, Robert L. Wilson Associates, Inc., Architects/Planner, Stamford, Connecticut

James R. Wright, Deputy Director, National Engineering Laboratory, National Bureau of Standards, Washington, D.C.

Joseph H. Zettel, Chairman, Building Research Advisory Board, Washington, D.C.

APPENDIX B

BIOGRAPHIES OF SPEAKERS

JOJI ARAI, Manager, U.S. Office, Japan Productivity Center, Washington, D.C. Mr. Arai has worked for the Japan Productivity Center since 1962. He served as program coordinator and assistant manager before assuming his present post. Earlier he served on the staff of the Judge Advocate, U.S. Air Force, and worked for the Agency for International Development. Mr. Arai studied law at Keio University.

WILLIAM H. BAHRKE, President, Dravo Utility Constructors, Inc., New York, New York. As chief executive officer of Dravo, Mr. Bahrke is responsible for construction, operation, field engineering, and partial design of power plants, industrial steam generator complexes, petrochemical plants, oil terminals, and the SPR project for the Department of Energy. Prior to joining Dravo, Mr. Bahrke spent over 20 years with Combustion Engineering, Inc., working on various aspects of power systems services (manufacturing, engineering, purchasing, project management, construction, and maintenance parts and centers). Mr. Bahrke is a member of the American Society of Mechanical Engineers, the Technical Association of the Pulp and Paper Industry, Inc., and the American National Standards Institute. He received a B.S. in mechanical engineering from Marquette University.

SEYMOUR A. BORTZ, Senior Engineering Advisor, Materials Technology Division, Construction Productivity Center, Illinois Institute of Technology (IIT), Chicago. Mr. Bortz has been conducting and managing research programs at the IIT Research Center since 1953. He has participated in projects involving the measurement and analysis of the mechanical properties of brittle materials including a wide range of construction materials. He is a member of the American Society of Civil Engineers, American Ceramic Society, American Society for Testing and Materials, American Concrete Institute, and Society for Experimental Stress Analysis. He is a registered professional engineer and holds B.S. and M.S. degrees from IIT.

JAMES M. BROWN, Professor of Law, National Law Center, George Washington University, Washington, D.C. Dr. Brown has held his current position at the university for the past 11 years and also serves as director of the Land Use Management and Control Program and as senior faculty participant in the university's Program of Policy Studies in Science and Technology. Prior to joining the George Washington University faculty, he taught law at the University of Mississippi and operated his own land development and construction business for 13 years. Dr. Brown has served on numerous panels and advisory committees of the National Academy of Sciences and the National Science Foundation. He received a B.A. degree from the University of Illinois and a J.D. degree from the University of Florida and was a Sterling Fellow at Yale Law School.

WILLIAM F. CALDWELL, President, Caldwell Equity Corporation, Troy, Michigan. Mr. Caldwell's company is engaged in both housing and commercial construction including residential subdivisions, apartment complexes, industrial parks, large shopping centers, office buildings, and condominium developments in Detroit, in New York and New Jersey, and on the west coast of Florida. Mr. Caldwell has been honored for eight consecutive years as a "Building Giant" by Professional Builder magazine and has received numerous architectural awards. He presently is a trustee and secretary of the Urban Land Institute and chairman of its Residential Council. He received a B.S. degree from Syracuse University.

WILLIAM J. CURTIN, Senior Partner, Morgan, Lewis and Bockius, Washington, D.C. Mr. Curtin has served as a public member of the Administrative Conference of the United States (1968-72), as chairman of the American Bar Association's Special Committee on National Strikes in the Transportation Industries, and as chairman of the District of Columbia Bar Association's Committee on Labor Relations Law. He presently is a Fellow of the American Bar Association and is chairman of the Labor Relations Law Committee of the Association's Public Utility Section. He received the American Arbitration Association's Award for Labor Management Peace in 1966 and Georgetown University's John Carroll Award in 1973. Mr. Curtin is a member of the Bar of the United States Supreme Court and holds LL.B. and LL.M. degrees from Georgetown University Law Center.

CHARLES J. DINEZIO, Executive Director, State Building Code Commission, Boston, Massachusetts. Mr. Dinezio served in his present capacity since 1973 and has pioneered the development of the statewide building code. He also served as coordinator of the Office of Code Development and as supervisor of training programs for code enforcement officials in the Massachusetts' Department of Community Affairs and as chief rehabilitation officer of the Charlestown Project for the Boston Redevelopment Authority. He currently is chairman of the Board of Trustees of the National Academy of Code Officials and immediate past president of the National Conference of State Building Codes and Standards. Mr. Dinezio also is a member of the Building Officials and Code Administrators International, Inc., International Congress of Building Officials, Southern Building Code Congress, Inc., and National Fire Protection Association.

THOMAS J. FLYNN, JR., Senior Vice President, Real Estate Loan Department, Bank of America (BoA), San Francisco, California. Prior to assuming his current position, Mr. Flynn was president of BankAmerica Realty Services, Inc.; vice president and head of BoA's Real Estate Loan Department (Northern Division); and vice president and BoA project manager for a $190 million southern California headquarters project (Atlantic Richfield Plaza). In the late 1960s, as president of California Land Company and vice president of Newhall Land and Farming Company, he was responsible for development of the Valencia new town community in southern California. He currently is a director of BankAmerica Realty Services, Inc.; BankAmerica Mortgage and International Realty, Inc.; Western America Financial, Inc.; and the Real Estate Research Council. He is a member of the Mortgage Bankers Association of America and the Foundation for Preservation of San Francisco's Architectural Heritage. Mr. Flynn holds a B.A. degree from and attended law school at the University of Santa Clara.

ROBERT A. GEORGINE, President, Building and Construction Trades Department, AFL/CIO, Washington, D.C. Mr. Georgine was elected secretary-treasurer of the Building and Construction Trades Department in 1971 and three years later was unanimously elected to his present post. Mr. Georgine started his career as an apprentice in Lathers' Local Union No. 74 in Chicago and has served as assistant business manager of the Lathing Foundation and international representative and president of the Wood, Wire and Metal Lathers' International Union. A recognized labor leader, Mr. Georgine is a member of the boards of the National Institute of Building Sciences, the American Productivity Center, the Atomic Industrial Forum, the American Nuclear Energy Council, the Advisory Council of the Electric Power Research Institute, the Gas Research Institute, the National Housing Conference, and the Alliance to Save Energy. Mr. Georgine attended the University of Illinois and DePaul University.

RICHARD P. GODWIN, Executive Vice President and Director, Bechtel Incorporated, San Francisco, Calfornia. Mr. Godwin joined Bechtel in 1961 as manager of research and development and, after serving in various management positions, assumed his current responsibilities as executive officer for the company's mining, civil, and nuclear fuels engineering and construction activities. Prior to joining Bechtel, Mr. Godwin served with the Atomic Energy Commission in positions dealing with plutonium production, nuclear weapons development, and reactor development and with the Maritime Administration as manager of research and engineering where he directed the building of the first nuclear merchant ship. Mr. Godwin is a member of the American Nuclear Society, the Society of Naval Architects and Marine Engineering, and the Yale Engineering Society. Mr. Godwin has an engineering degree from Yale University and is a registered professional nuclear engineer.

JOHN W. GUINEE, JR., Real Estate Investor, Reston, Virginia. Mr. Guinee formerly was president of the Yeonas Company, Vienna, Virginia. Before joining Yeonas, he was executive vice president of Gulf-Reston and the Gulf-Reston Real Estate Corporation and was responsible for Gulf-Reston's activities in Virginia, Florida, Kansas, and Puerto Rico. He also headed his own building firm and worked as a management consultant in the United States and Latin America for such builders as Levitt and Sons.

G. RALPH GUTHRIE, President, Urban Investment and Development Company, Chicago, Illinois. Urban Investment, one of the nation's largest real estate and development firms, is engaged in construction of single-family and multifamily residential properties, general contracting for retail and high-rise construction, and real estate investment, development and management of various multiple-use structures and facilities. Prior to joining Urban, Mr. Guthrie served in key executive positions with the N. K. Winston Corporation and I-T-E Imperial Corporation. Mr. Guthrie is a member of the Financial Planning Council of the American Management Association, the Financial Executives Institute, the Urban Land Institute, the Cosmopolitan Chamber of Commerce, the Chicago Association of Commerce and Industry, the Economic Club, and the Executives Club. He received a B.S. degree in economics from the Wharton School of the University of Pennsylvania.

JEFFREY J. HALLETT, Director, Productivity Center, Chamber of Commerce of the United States, Washington, D.C. Mr. Hallett has been associated with productivity analysis and program development for the past seven years. He has served as an assistant director of the National Center for Productivity and Quality of Working Life and as a policy advisor to assistant secretaries in both the U.S. Departments of Commerce and of Health, Education and Welfare. He also has started and managed three small businesses and has been involved in real estate syndications for inner city townhouse restorations. He currently serves on the board of two corporations in Washington that are providing services and research on the management of change in both public and private institutions. Mr. Hallett received a B.A. degree from Wesleyan University and an M.B.A. degree from Harvard University.

CALVIN S. HAMILTON, Director of Planning, Department of City Planning, Los Angeles, California. Mr. Hamilton is responsible for all proposed zoning regulations and requirements for Los Angeles and investigates and reports on the design and improvement of all proposed subdivisions of land. Before assuming his present post in 1964, Mr. Hamilton was executive director of the Pittsburgh Department of City Planning. He is a member of the Building Research Advisory Board, American Institute of Planners, American Society of Landscape Architects, and International Federation of Housing and Planning. He holds a bachelor of fine arts degree from the University of Illinois and a master of city planning degree from Harvard University and was a research associate of University College, Department of Town Planning, University of London.

L. KENNETH HARMON, Corporate Productivity Coordinator, Mead Corporation, Dayton, Ohio. In his current position, Mr. Harmon's chief responsibility is to strengthen productivity management within the various divisions of the corporation. He joined the corporation in 1972 and has served as manager of operations analysis and manager of internal auditing and represented the corporation as associate member on the staff of the American Productivity Center. Earlier Mr. Harmon was a management and marketing consultant, a research associate and management instructor at Ohio State University, and an industrial engineer for the U.S. Air Force's Construction and Maintenance Division. He is a past member of the board of the American Institute of Industrial Engineers and a member of the Engineers Club of Dayton. He received a B.S. degree in industrial engineering from Georgia Tech and a M.B.A. degree from Wright State University.

ROBERT C. HOLLAND, President, Committee for Economic Development, Washington, D.C. Since 1976 Dr. Holland has been executive officer for the Committee for Economic Development (CED), a non-profit research and educational organization of business and academic leaders that is devoted to the study of public policy problems. He was associated with the Federal Reserve System's Board of Governors from 1964 to 1976 and served as a member of the Board from 1973 to 1976. Earlier he held various research and administrative staff positions including executive director, secretary of the board, and secretary of the Federal Open Market Committee. He also was an instructor in money and banking at the Wharton School of the University of Pennsylvania and was employed at the Federal Reserve Bank of Chicago in various positions ranging from financial economist to vice president. He is a member of the Council on Foreign Relations, the United Nations Association's Business and Labor Economic Policy Council, the Wharton School Board of Overseers and the Wharton School Finance Department Advisory Committee, the Lutheran Resources Commission Advisory Committee, and the Comptroller General's Consultant Panel. Dr. Holland holds a B.S. degree in finance, an M.A. degree, and a Ph.D. degree in economics from the University of Pennsylvania and was awarded an honorary doctor of laws degree by the University of Nebraska.

RONALD E. JEANES, Head, Planning and Construction Department, Building Research Establishment (BRE), Garston, Watford, England. Mr. Jeanes has been affiliated with the BRE since 1962 and has served in various posts with responsibility for research program formulation in the fields of housing and building; support services, administration, and application of research; and technical and management services. He also led a BRE operational research team concerned with productivity studies, network planning, and the study of operative skills. After graduating in mathematics from University College, Exeter, he served in the Royal Naval Scientific Service and worked in operational research concerning underwater weapon detection and guidance systems.

OLIVER H. JONES, Consulting Economist, Oliver Jones and Associates, Manns Choice, Pennsylvania. Dr. Jones organized his firm after retiring as executive vice president of the Mortgage Bankers Association of America. Earlier he had worked with the Federal Reserve System, the University of California, and the Sanford Research Institute. Dr. Jones is a member of the Building Research Advisory Board, American Statistical Association, American Economics Association, American Finance Association, and National Association of Business Economists. He holds M.A. and Ph.D. degrees from Pennsylvania State University.

F. S. KELLSTROM, Chairman of the Board, Fishbach and Moore, Inc., Los Angeles, California. Mr. Kellstrom serves as chief executive officer of one of the largest national and international electro-mechanical contracting firms. He has been with the organization since 1946 and served as vice president, executive vice president, and president before becoming chairman. He is a member of the National Electrical Contractors Association and past president of the Los Angeles Chapter and has served on many of the Association's committees.

CHARLES B. KNAPP, Deputy Assistant Secretary of Labor for Employment and Training, U.S. Department of Labor, Washington, D.C. Dr. Knapp is responsible for implementation of several federal government programs designed to assist individuals who are disadvantaged in the labor market including the Comprehensive Employment and Training Act. Dr. Knapp joined the Department in 1977 as a special assistant to the Secretary of Labor and was responsible for program development and other special projects. Formerly, he had been an assistant professor of economics and a research associate at the Center for the Study of Human Resources, University of Texas. He also served as consultant to various organizations including the Ford Foundation and the Rand Corporation. He received a B.S. degree from Iowa State University and M.A. and Ph.D. degrees from the University of Wisconsin.

CREIGHTON C. LEDERER, Commissioner and Director, Buildings and Safety Engineering Department, Detroit, Michigan. Mr. Lederer has worked in various engineering positions for Detroit for over 25 years and was chief bridge engineer in the City Engineers Office prior to assuming his current position. Earlier he had worked as a bridge designer for the New York Central Railroad. Mr. Lederer has served on the executive committees of the Building Officials and Code Administrators International, Inc. (BOCA) and the American Major City Building Officials (AMCBO) and as vice chairman of the Michigan State Construction Code Commission. He is a Fellow of the American Society of Civil Engineers and the Engineering Society of Detroit and is a member of the American Railway Bridge and Building Association, the Society of Municipal Engineers, the American Public Works Association, and the National Academy of Code Administrators. He is a registered professional engineer and received a B.S. degree in civil engineering from the University of Michigan.

JOHN W. LEONARD, Vice President, Morrison-Knudsen Company, Inc., Boise, Idaho. Mr. Leonard has been affiliated with Morrison-Knudsen for over 30 years and worked directly on projects before assuming an executive position. He is a member of the American Arbitration Association, American Society of Military Engineers, Beavers, and Moles and is a Fellow of the American Society of Civil Engineers. He is a registered engineer and is a graduate of the Massachusetts Institute of Technology.

WALTER H. LEWIS, Professor, Department of Architecture, University of Illinois, Urbana. Professor Lewis has served on the faculty of the Department of Architecture for the past 20 years. He has gained national and international recognition for his work in the fields of housing, construction, technology, regulation, code administration and enforcement, and the environmental aspects of town and community planning. He is a charter incorporator of the National Academy of Code Administration and currently serves on its board of directors and executive committee. Professor Lewis has received national awards for creating continuing professional educational programs for construction industry personnel, has served on numerous committees for the American Institute of Architects, and is environmental consultant to the U.S. League of Savings Associations (which represents 95 percent of the savings and loan business). Prior to his academic career, he was in private architectural practice. He received a B.A. degree in architecture from the University of Illinois.

ALBERT RHOADES MARSCHALL, Commissioner, Public Buildings Service, General Services Administration, Washington, D.C. As commissioner of the Public Buildings Service, one of the nation's largest civilian real estate operations, Adm. Marschall is responsible for managing approximately 10,000 buildings comprising some 232 million square feet and an annual operating budget of $1.4 billion. During his 33-year career with the U.S. Navy, he served as commander of the U.S. Naval Engineering Command (NAVFAC), chief of engineers, commanding officer of the NAVFAC Southern Division, officer in charge of construction in Vietnam, and director of the Navy's facilities planning and programming system. He also served as vice president of the George Hyman Construction Company and managed $1.5 billion in facilities construction. Adm. Marschall is a member of the National Society of Professional Engineers, the American Society of Civil Engineers, and the Moles. He is past president of the Society of American Military Engineers and past director of the American Public Works Association. A graduate of the U.S. Naval Academy, he received B.S. and M.S. degrees in construction engineering from Rensselaer Polytechnic Institute.

ROBERT P. MARSHALL, JR., Vice Chairman, Turner Construction Company, New York, New York. Mr. Marshall has been employed by Turner for the past 40 years (with exception of active service in World War II). He has served in various field positions with responsibility for management of the construction of institutional, industrial, and commercial buildings. In his current position, he is the administrative director of all corporate activity relating to sales, marketing, contracts, personnel, public relations, advertising, legal matters, estimating, and purchasing. Throughout his career he has actively served local and national contractors' associations (as director, committee chairman, and committeeman) and has been involved especially in activities related to manpower, labor relations, contractual relations, and legislation. He is a member of the American Arbitration Association (Arbitor), the American Society of Civil Engineers, the Associate General Contractors of America, and the Society of American Military Engineers. Mr. Marshall received a B.S. degree in civil engineering from the University of Pennsylvania.

NEIL B. McARTHUR, Vice President, The Austin Company, Washington, D.C. Mr. McArthur has been with Austin for over 10 years and is responbile for the company's domestic labor relations. He came to the company with more than 20 years of experience in labor relations, having served as president and business agent for the Detroit Carpenter's Local Union, as commissioner of the Michigan Department of Labor, and as special assistant to the Assistant Secretary for Manpower of the U.S. Department of Labor. Mr. McArthur is a member of the Board of Directors and Labor Relations Council of the National Constructors Association and past chairman of its Labor Relations Committee. He also is a member of the International Union of Operating Engineers' Apprenticeship Council, served as Manpower Advisor to the Equal Employment Opportunities Commission, and established the first Outreach program for the building trades.

DAVID S. MILLER, President, David S. Miller and Associates, Inc., Cleveland, Ohio. Mr. Miller's firm provides consulting services to the building industry on housing technology, market and product research, marketing, and management. He is a member of the Board of Directors of the Producers' Council, the Policy Advisory Board to the Harvard-MIT Joint Center for Urban Studies, and the Executive Committee of the U.S. Chamber of Commerce's Construction Action Council. He served on the HUD Task Force on Housing Costs (1977-78) and is past president of the Producers' Council and immediate past chairman of the National Institute of Building Sciences.

CHARLES E. "Ted" PECK, Executive Vice President, Owens-Corning Fiberglas Corporation, Toledo, Ohio. Mr. Peck joined Owens-Corning in 1949 and has held a variety of managerial positions including vice president and general manager, Home Building Products Division, and group vice president, Construction Group. He was elected to the Board of Directors in December 1975. In the Office of the Chief Executive, Mr. Peck has oversight responsibility for the Supply Division and the Insulation, Roofing Products, and Interiors Operating Divisions. Mr. Peck is a member of the Board of Directors of the U.S. Chamber of Commerce; chairman of the Executive Committee of the U.S. Chamber's

Construction Action Council, which advises the Chamber on issues related to the construction industry; and chairman of the Producers' Advisory Forum. He is a director of The Ryland Group, a multicity home building company. He also is vice chairman of the Visiting Committee of the MIT-Harvard Joint Center for Urban Studies and a past member of the National Bureau of Standards' Statutory Visiting Committee and the Federal National Mortgage Association Advisory Board. Mr. Peck holds a B.S. degree from the Wharton School of the University of Pennsylvania and serves on the Advisory Board of Mercy Hospital in Toledo.

R. THAYNE ROBSON, Director, Bureau of Economic and Business Research, and Professor of Management and Economics, University of Utah, Salt Lake City. Mr. Robson has been affiliated with the school for 12 years and has served as associate dean of the College of Business. He also has taught economics, industrial relations, management, and urban policy at the University of California and the Brookings Institution and has worked with the U.S. Departments of Labor, Housing and Urban Development, and Health, Education, and Welfare. He has served in such positions as senior staff economist for the National Commission on Technology, Automation and Economic Progress, executive director of the President's Committee on Manpower, chairman of the Select Task Force on Recruitment, Referral and Placement of the JOBS program, and chairman of the Utah Manpower Planning Council. He is a member of the Utah State Advisory Council for Vocational and Technical Education and the Utah Consortium for Energy Research and Education. He received B.S. and M.S. degrees in economics from the Utah State University and is a doctoral candidate in industrial relations and economics at Cornell University.

JEROME L. ROSENBERGER, Construction Manager, Hydrocarbon Engineering Division, Union Carbide Corporation, Houston, Texas. Mr. Rosenberger manages Union Carbide's Construction Group, which supports the capital construction of seven of the corporation's operating divisions, and currently is concentrating on developing construction technology and a construction management system. His experience in the management of major and minor construction includes the building of petrochemical plants for Monsanto Chemical Company. Mr. Rosenberger is a member of study teams on construction technology and construction management of the National Business Roundtable's Cost-Effectiveness Task Force and of the American Association of Cost Engineers. He received a B.S. degree in civil engineering from the University of Alaska and is a registered professional engineer.

ROBERT F. SCHMITT, President, Bob Schmitt Homes, Inc., Strongsville, Ohio. Mr. Schmitt has been a residential builder and land developer for more than 20 years and has built more than 2000 homes and condominium units. He is active in building regulatory matters on the national, state, and local levels and is recognized as a national leader in the field of energy conservation in residential building.

Mr. Schmitt serves as chairman of the National Association of Home Builders (NAHB) Research Foundation and is a member of the Board of Directors of the National Institute of Building Sciences. He has been named "Builder of the Year" by Professional Builder magazine and has been included into the NAHB Hall of Fame. Mr. Schmitt received a B.S. degree in industrial engineering from the Ohio State University.

IRVING H. SIEGEL, Consulting Economist, Bethesda, Maryland. Dr. Siegel retired recently as economic adviser to the Bureau of Domestic Business, Industry and Trade Administration, U.S. Department of Commerce, and has served as a consulting economist to the American Chemical Society; the Patent, Trademark and Copyright Research Institute of the George Washington University, where he specialized in the economic and social aspects of technical change, and the IBM Corporation, where he designed a productivity measurement system. Before joining the Department of Commerce, he was an economist for the W. E. Upjohn Institute for Employment Research, director of the Soviet Productivity Study for the Research Analysis Corporation, director of the Johns Hopkins University Operations Research Office, senior economist for the President's Council of Economic Advisors, director of the American Technology Study for the Twentieth Century, and chief economist for the Veterans Administration. Dr. Siegel has authored numerous works and has lectured extensively on a wide range of economic, statistical, technological, and managerial subjects. He is a Fellow of the American Statistical Association, the New York Academy of Sciences, and the American Association for the Advancement of Sciences. He holds a B.S. degree from the City College of New York, an M.A. degree from New York University, and a Ph.D. from Columbia University.

EDWARD L. SIMONS, Manager, Environmental Protection Operation, General Electric Company, Schenectady, New York. As manager of GE's Environmental Protection Operation (EPO), Dr. Simons heads the organization responsible for monitoring, counseling, appraising, and reporting on the environmental protection programs at the more than 200 GE plants and facilities throughout the world. Prior to assuming his present position, Dr. Simons was manager of the EPO's Environmental Information Center, which he established after joining the EPO in 1971. In that position, he was responsible for providing a mechanism by which the large amount of environmental information and data that reaches the EPO from GE, government, and outside sources can be evaluated, interpreted, and organized into packages of pertinent information that will be useful to GE personnel concerned about particular aspects of environmental protection. Previously, Dr. Simons had served for 20 years at the GE Research and Development Center carrying out research and consulting activities in analytical chemistry, corrosion, and electrochemistry. He also managed the Center's fuel cells programs and served as the full-time staff member on GE's Corporate Environmental Task Force. Before joining GE, he was a research chemist on the Manhattan project and a chemistry professor at Rutgers University. Dr. Simons is a member of the American Association for the Advancement of Science and the American Chemical Society. He received a B.S. degree from the City College of New York and a Ph.D. from New York University.

DENNIS J. SULLIVAN, Vice President for Government Services, American Productivity Center, Inc., Houston, Texas. Mr. Sullivan is responsible for all of the Center's productivity measurement and improvement work involving government organizations and government activities that have an impact on the productivity of private sector organizations. Prior to joining the Center's staff, he spent 15 years as a consultant specializing in individual and organizational performance assessment and training system development in the United States and abroad. Mr. Sullivan holds a graduate degree from the University of Southern California.

VANCE W. TORBERT, JR., Architectural Officer, Real Estate Financing, Metropolitan Life Insurance Company, New York, New York. Mr. Torbert joined Metropolitan in 1955 as city loan architect and architectural officer in the company's Real Estate Financing Department. Prior to that he was engaged in private professional practice with Voorhees, Walker, Foley and Smith, Architects, and Giffels and Vallet, Architects-Engineers. He has served on the National Bureau of Standards' Evaluation Panel for Energy Conservation Programs and participated in the Harvard University conference on the real cost of building. Mr. Torbert received a B.A. degree in architecture from Princeton University and studied advanced construction at New York Structural Institute.

MURRAY L. WEIDENBAUM, Director, Center for the Study of American Business, Washington University, St. Louis, Missouri, currently is on leave serving as a resident scholar at the American Enterprise Institute, Washington, D.C. He joined Washington University in 1964, served as chairman of the Economics Department, and has held the Mallinckrodt Distinguished University Professorship since 1971. Formerly, Dr. Weidenbaum served as Assistant Secretary of the Treasury for Economic Policy, a fiscal economist in the U.S. Bureau of the Budget, and the corporate economist at the Boeing Company. He is the consulting economist to the First National Bank in St. Louis and to Mallinckrodt, Inc. Dr. Weidenbaum is a Fellow in the National Association of Business Economists and served on the original Board of Directors of the National Economist Club. He also is a member of the Council on Foreign Relations and the Board of Economists of Time magazine. He has been awarded the Townsend Harris Medal by the City College Alumni Association, the Distinguished Writers Award by the Center for Strategic and International Studies, and the Alexander Hamilton Medal "in recognition of distinguished leadership in the Department of the Treasury." Dr. Weidenbaum received a B.S. degree from the City College of New York, an M.A. from Columbia University, a Ph.D. in economics from Princeton University.

ROBERT M. WHITE, Administrator, National Research Council, and Executive Officer, National Academy of Sciences, Washington, D.C. Before assuming his present positions, Dr. White was president of the Joint Oceanographic Institutions, Inc., and chairman of the National Academy of Sciences' Climate Research Board. Earlier he served as administrator of the National Oceanic and Atmospheric Administration and of the U.S. Department of Commerce's Environmental Science Services Administration. The recipient of many awards and honors, Dr. White has served on a variety national committees, commissions, and boards. He received his Sc.D. degree from the Massachusetts Institute of Technology.

ROBERT L. WILSON, AIA, President, Robert L. Wilson Associates, Inc., Architects/Planners, Stamford, Connecticut. Mr. Wilson has been in private practice since 1966 designing commercial, institutional, residential, and transportation facilities and serving as a management consultant on plant operation. He is national vice president and a member of the board of the American Institute of Architects and is co-founder and member of the National Organization of Minority Architects and New York Coalition of Black Architects. He holds a master of architecture degree from Columbia University.